YACHTMASTER EXERCISES

PAT LANGLEY-PRICE

AND

PHILIP OUVRY

ADLARD COLES NAUTICAL
London

Adlard Coles Nautical
an imprint of A & C Black (Publishers) Ltd
35 Bedford Row, London WC1R 4JH

First published in Great Britain by
Adlard Coles 1984
Reprinted 1987, 1988, 1989
Reprinted by Adlard Coles Nautical 1991

British Library Cataloguing in Publication Data

Langley-Price, Pat
Yachtmaster exercises.
1. Navigation — Problems, exercises etc.
2. Yachts and yachting — Problems, exercises,
etc.
I. Title   II. Ouvry, Philip
623.89'0247971   VK559.5

ISBN 0-229-11715-5

Printed and bound in Great Britain by
Mackays of Chatham PLC, Chatham, Kent

Contents

Acknowledgements

Thomas Reed Publications Ltd, extracts from *Reed's Nautical Almanac*.

Ministry of Defence, Hydrographic Dept, extracts from *Admiralty Tide Tables*, and Practice Chart 5055.

The Cruising Association, extracts from *Cruising Association Handbook*.

Adlard Coles Ltd, extracts from *Normandy and Channel Islands Pilot*, 5th edn (1983), Mark Brackenbury.

Introduction

This book of exercises has been prepared as a supplement to *YACHTMASTER An Examination Handbook with Exercises*. Navigation students have frequently found difficulty with tidal problems and chartwork. It has been assumed that the reader already has textbooks on small boat navigation and is seeking to extend his knowledge with additional exercises. Thus no theoretical explanations are included, though some answers are expanded, where necessary, to include further clarification. The papers are graded in difficulty. To enable the exercises to be completed, all relevant extracts from hydrographic and nautical publications have been included together with Instruction Chart 5055. The answers given are set out as model answers, and are very complete.

The extracts are from a variety of sources: *Admiralty Tide Tables NP201*, tidal information; *Reed's Nautical Almanac*, tidal information, pilotage, radio beacons, distance off tables; *Cruising Association Handbook*; and *Normandy and Channel Islands Pilot*, pilotage. Whilst these extracts are intended to be examples of the hydrographic information needed by yachtsmen, it is emphasised that an adequate range of publications and charts is required in practice. For instance; several examples use Rye harbour as a destination and this is a difficult harbour to enter without local knowledge. Published information generally available is barely sufficient to permit any but an experienced yachtsman to enter the harbour.

In many instances bearings of marine radio beacons have been used to fix positions. This has been done because of lack of suitable landmarks on the practice chart; a larger scale chart would include many more such landmarks. Use of radio bearings for accurate fixes, particularly at ranges in excess of 10 miles, is not generally considered good practice.

Accuracy achieved in theory is often difficult to achieve in real life. When doing any navigational exercises, readers should strive as much as possible for accuracy and tidy presentation. Much navigational information is inexact, for example: leeway, tidal height offshore, tidal streams in positions between tidal diamonds. Accurate navigation requires continued application of the best information available. More experienced navigators know how much to leave out, but the learner should seek information from as many sources as possible.

To conclude, remember that there is no substitute for practical experience. One week spent navigating a boat on an offshore passage is more valuable than a whole winter spent at shorebased theory classes. The authors, therefore, have set out these exercises hoping that reaching the end of this book will be a prelude to many years of enjoyable navigating.

Pat Langley-Price and Philip Ouvry

By the same authors:

YACHTMASTER   (Adlard Coles)
ISBN 0-229-11662-0

OCEAN YACHTMASTER   (Adlard Coles)
ISBN 0-229-11695-7

OCEAN YACHTMASTER EXERCISES   (Adlard Coles)
ISBN 0-229-11792-9

VHF YACHTMASTER   (Adlard Coles)
ISBN 0-229-11720-1

COMPETENT CREW   (Adlard Coles)
ISBN 0-229-11736-8

USING YOUR DECCA   (Adlard Coles)
ISBN 0-229-11853-4

Definitions

Chart Datum The reference level on charts. Soundings are given below chart datum. Heights of tide are heights above chart datum.

Course The direction in which the boat is heading. True (T), magnetic (M) and compass (C) courses are the angles between the boat's fore and aft line and the direction of true, magnetic and compass north respectively (measured clockwise from north).

Course To Steer The compass course to be steered by the helmsman. It is found by calculation or by plotting a vector diagram allowing for tidal stream, leeway, variation and deviation.

Depth The vertical distance between sea level and the seabed.

Distance Made Good The distance travelled over the ground taking into account the boat's course and speed, tidal stream and leeway.

Drying Height The height above chart datum of any feature which is covered and uncovered by the tide. Drying heights are underlined on the chart.

Duration Of Tide The period of time between high water and low water for a falling tide and low water and high water for a rising tide.

Ground Track The track made good over the ground allowing for the boat's course and speed, tidal stream and leeway.

Heading The direction of the boat's fore and aft line.

Height The heights of natural features above sea level are given in metres. In the case of a light the height is between the focal plane of the light and MHWS.

Height Of Tide The distance between sea level and chart datum at any instant.

Interval The period of time between any instant and the preceding or following HW or LW.

Leeway The sideways drift of a boat caused by the wind. Often expressed as the angle between the course and the water track, and most easily measured astern of the fore and aft line (the reciprocal of the boat's heading) and her wake. The leeway angle is normally between 5° and 10° when the boat is closehauled depending on the underwater profile of the boat. Bilge keeled boats make more leeway than fin keeled boats. Leeway decreases progressively as the boat bears away to a beam reach, becoming nil when she is running. Leeway has to be taken into account when determining a course to steer and when plotting a water track.

Mean High Water Springs (MHWS), Mean Low Water Springs (MLWS), Mean High Water Neaps (MHWN), Mean Low Water Neaps (MLWN) The mean throughout the year of the heights above chart datum of all HW and LW spring and neap tides.

Range The difference in metres between the height of HW and the preceding or succeeding LW.

Rise Of Tide At any instant the distance of sea level above LW.

Sounding The (charted) depth of the seabed below chart datum.

Speed Made Good The speed made good over the ground taking into account the boat's course and speed, tidal stream and leeway.

Tidal Stream

Set The direction to which a tidal stream or current flows.

Rate The speed at which a tidal stream runs, given in knots and tenths of a knot.

Drift The distance that a boat is carried by a tidal stream or current in a given time when no other factors are considered.

Track Made Good The mean track actually achieved over the ground by a boat during a given period of time.

Water Track The track travelled through the water allowing for the effect of leeway (but not tidal stream).

Symbols and Abbreviations

+	Dead reckoning position (DR). Position derived only from course steered and distance travelled.
△	Estimated position (EP). Position derived from course steered, distance travelled, leeway and tidal stream.
⊙	Fix. Position derived by direct reference to terrestrial landmarks.
→	Water track.
⇉	Ground track.
⇶	Tidal Stream.
→	Position line.
	Position obtained from a bearing and a position circle.
	Position obtained from two position circles.
⇇ ⇉	Transferred position line.
BST	British summer time (Greenwich mean time plus one hour).
C	Compass bearing or course.
CD	Chart datum.
CE	Compass error (the sum of variation and deviation).
E	East.
ETA	Estimated time of arrival.
GMT	Greenwich mean time.
h	Hours.
HW	High water.
LW	Low water.

M   Magnetic bearing or course, or miles.
m   Metres or minutes
MHWS   Mean high water springs.
N   North.
RC   Marine radio beacon.
RDF   Radio direction finder.
S   South.
s   Seconds.
T   True bearing or course.
TSS   Traffic separation scheme.
VHF   Very high frequency (radio).
W   West.

Notes on Exercises

1. All times shown are British Summer Time (BST) unless stated.
2. Variation is 4° west throughout.
3. Deviation taken from Deviation Table, Fig. 3.1 (page 4) is to the nearest whole degree.
4. Approximations are made to the nearest tenth of a knot, tenth of a metre in height or depth, tenth of a nautical mile and to the nearest minute in time and degree of angle.
5. Practice chart 5055 has been updated and the new chart is included with this book.
6. Soundings are, by definition, corrected for height of tide.
7. Buoys are not usually visible at distances of over one mile and should be used for bearings only if no suitable landmarks are available.
8. In several instances, courses to steer and estimated time of arrival are calculated to a harbour breakwater rather than to the centre of the harbour.

CHAPTER ONE

Position and Direction

1.1 Give the latitude and longitude of the following:

(a) The RW buoy in Rye Bay.
(b) Royal Sovereign lighthouse.
(c) Varne light vessel.
(d) ZC2 buoy (north west of Cap Gris-Nez).

1.2 What symbols are printed on chart 5055 in the following positions?

(a) 50°32′.8N, 0°57′.8E.
(b) 50°26′.9N, 1°00′.1E.
(c) 50°35′.7N, 1°19′.8E.
(d) 50°44′.9N, 1°27′.2E.

1.3 What is the true bearing of Bassurelle light buoy from Vergoyer SW cardinal buoy?

1.4 A boat is 3 miles west of Cap d'Alprech light (Fl (3) 15s). What is the latitude and longitude of its position?

1.5 Boulogne south breakwater light (Fl (2+1) 15s) bears 074°T distance 2 miles. What is the latitude and longitude of the boat's position?

CHAPTER TWO

Buoys and Lights

2.1 What are the following types of buoys and what do they indicate:

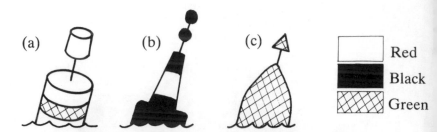

2.2 (a) What type of buoy is in position 50°26′.9N, 1°00′.1E?
(b) What does it indicate?

2.3 Describe the characteristics of Cap d'Alprech light (south of Boulogne).

2.4 (a) Describe the light structure in position 50°52′.0N, 1°35.′0E (as seen by day). (See Extract 18).
(b) Between which bearings from seaward is the light visible?

2.5 (a) What type of mark is ZC1 buoy (4.5 miles west of Boulogne)?
(b) What is its purpose?

2.6 (a) How could Beachy Head lighthouse be recognised in fog?
(b) Could it be confused with Royal Sovereign lighthouse?

CHAPTER THREE

Variation and Deviation

Variation is as shown.
Use the deviation table Fig 3.1 except wh e deviation is given.

3.1 Convert the following magnetic be   gs to true bearings:

	Bearing	Variation
(a)	169°M	6°E
(b)	205°M	7°W
(c)	094°M	3°W
(d)	358°M	9°E

3.2 Convert the following to magnetic bearings:

	Bearing	Deviation	Variation
(a)	168°C	8°W	
(b)	304°C	11°E	
(c)	047°T		4°E
(d)	291°T		5°W

3.3 Convert the following to true bearings using Fig 3.1 and Variation 6°W:

(a) 330°C
(b) 241°C
(c) 058°C
(d) 162°C

3.4 A boat steering a course of 223°C takes the following bearings using the steering compass. Convert these bearings to true. Variation 8°W.

Church	358°C
Tower	265°C
Lighthouse	134°C

3

3.5 Find the compass error (CE) and the deviation from the following bearings (do not use the deviation table). Variation 7°E.

	True bearing	Compass bearing
(a)	194°T	199°C
(b)	216°T	210°C
(c)	355°T	007°C
(d)	001°T	353°C

Fig. 3.1   Deviation Table

Compass heading	Deviation
000°	5°E
020°	5°E
040°	$4\frac{1}{2}$°E
060°	4°E
080°	$3\frac{1}{2}$°E
100°	$1\frac{1}{2}$°W
120°	$2\frac{1}{2}$°W
140°	4°W
160°	5°W
180°	$4\frac{1}{2}$°W
200°	$3\frac{1}{2}$°W
220°	2°W
240°	0°
260°	$2\frac{1}{2}$°E
280°	4°E
300°	$4\frac{1}{2}$°E
320°	5°E
340°	5°E

CHAPTER FOUR

Estimated Position

4.1 At 1430 a boat is in position 50°23′.4N, 1°23′.6E. Her course is 010°T, and her speed 5.5 knots. Leeway is negligible and the tidal stream is slack.

What is the EP at 1530? –

4.2 At 1750 a boat is in position 50°30′.3N, 1°32′.2E steering a course of 195°T, at a speed of 4 knots. Leeway is negligible.
Tidal streams:

1750 – 1850	202°T 1.7 knots
1850 – 1950	197°T 1.6 knots

What is the EP at 1950?

4.3 At 0840 a boat is in position 50°33′.7N, 1°20′.0E steering a course of 126°T, at a speed of 3.5 knots. The wind is SW 3, leeway is 5°.
Tidal streams:

0840 – 0940	031°T 1.5 knots
0940 – 1040	020°T 1.6 knots

(a) What is the EP at 1040?
Between 0840 and 1040:
(b) What is the water track?
(c) What is the ground track?
(d) What is the distance made good?
(e) What is the speed made good?

4.4 At 1430 a boat is in position 50°19′.1N, 1°16′.5E steering a course of 334°M, at a speed of 6 knots. The wind is NE 5, leeway is 10°.
Tidal streams:

1430 – 1530	048°T 1.2 knots

5

1530 – 1630   030°T 1.0 knot

What is the EP at 1630?

4.5   At 0929, log 58.4, a boat is in position 50°32'.2N, 1°21'.8E steering a course of 188°M. The wind is E 4, leeway is 5°.
Tidal streams:

0929 – 1029   021°T 1.6 knots
1029 – 1129   021°T 1.4 knots

What is the EP at 1129 when the log reads 67.3?

4.6   The following is an extract from a boat's log:

Time	Course	Log	Wind	Leeway	Remarks
2015	120°T	124.1	S 4	10°	Position 50°24'.2N, 1°11'.8E
2045	130°T	126.5	S 4	10°	
2200		132.8			

Tidal streams:

2000 – 2100   042°T 0.8 knots
2100 – 2200   032°T 1.4 knots

What is the EP at 2200?

4.7   At 0200, log 10.9, a boat steering a course of 314°M, fixes her position as 270°T Pt du Lornel light (Oc (2) WRG 6s) 2.1 miles. At 0330, log 18.4 she tacks to a course of 054°M. The wind is N 4, leeway is 10°.
Tidal streams:

0200 – 0300   011°T 1.1 knots
0300 – 0400   014°T 0.9 knot

What is the EP at 0400, when the log reads 20.9?

4.8   At 1617, log 96.4, a boat is in position 50°26'.1N, 1°11'.6E. She is close hauled on port tack steering a course of 024°M. At 1717, log 101.5, she tacks to a course of 294°M. At 1817 the log reading is 106.4. Leeway is 10° on both tacks.
Tidal streams:

1617 – 1717   082°T 1.0 knot
1717 – 1817   057°T 1.2 knots

(a) What is the EP at 1717?
(b) What is the EP at 1817?

4.9   At 1145, log 58.2, a boat is in position 50°21'.6N, 1°04'.6E
She is on port tack steering a course of 320°T. At 1243, log 64.2,
she tacks through 90°. Leeway on both tacks is 10°.
Tidal streams:

1145 – 1245	278°T 0.6 knots
1245 – 1345	247°T 0.9 knots
1345 – 1445	235°T 1.2 knots

(a) What is the EP at 1400, log 73.4?

(b) What is the speed made good between 1145 and 1400?

4.10 At 1920 a boat in position 50°18'.1N, 0°50'.2E is close hauled
on starboard tack steering a course of 049°M. At 2050 she tacks
through 90°, and at 2220 tacks again to a course of 032°M. The
boat's speed is 5 knots. Leeway is 10° throughout.
Tidal streams:

1920 – 2020	082°T 1.0 knot
2020 – 2120	057°T 1.2 knots
2120 – 2220	048°T 1.2 knots
2220 – 2320	030°T 1.0 knot

What is the EP at 2250?

CHAPTER FIVE

Course to Steer

5.1 At 1215 a boat is in position 50°41′.9N, 0°00′.3E. Her speed is 3 knots. Leeway is negligible and the tidal stream is slack:

    (a) What is the true course to Newhaven Breakwater lighthouse?

    (b) What is the ETA at the breakwater?

5.2 At 1650 a boat is in a position 180°T Royal Sovereign lighthouse 1.9 miles. The boat's speed is 5.5 knots. Leeway is negligible. Tidal streams:

        1650 – 1750   249°T 1.4 knots
        1750 – 1850   248°T 1.8 knots

    (a) What is the magnetic course to position 50°43′.7N, 0°07′.8E?

    (b) What is the ETA at this position?

5.3 At 1020 a boat is in position 50°37′.5N, 0°26′.5E. The boat's speed is 5.6 knots. Leeway due to a westerly wind is 5°. Deviation is 4°E.

Tidal streams:

        1020 – 1120   068°T 1.9 knots
        1120 – 1220   068°T 2.6 knots

    (a) What is the course to steer to position 50°47′.6N, 0°32′.5E?

    (b) What is the ETA at this position?

5.4 At 1015 a boat fixes her position as 50°40′.9N, 0°35′.4E. The boat's speed is 6 knots. The wind is ESE 5, leeway is 10°.
Tidal stream:

        1015 – 1115   068°T 2.6 knots

(a) What is the magnetic course to position 50°47′.1N, 0°45′.4E?
(b) When will the boat arrive at this position?
(c) Which tack is the boat on?

5.5 At 0630, a boat in position 50°48′.3N, 0°40′.6E wishes to reach the RW buoy in Rye Bay. She can sail 50° off the wind on either tack maintaining a speed of 4.5 knots. The wind is due north, leeway is 10° and the tidal stream is negligible.

(a) What is the true course on each tack?
(b) Which should be the initial tack?
(c) At what time should the boat tack?
(d) What is the ETA at the buoy?

5.6 At 0400 a boat in position 50°58′.2N, 1°14′.0E wishes to reach Hythe buoy, south west of Folkestone. She can sail 270°T on starboard tack and she can tack through 90°. The boat's speed is 4 knots. The wind is NW 3, leeway is 10°.
Tidal streams throughout:
    058°T   2.6 knots

(a) Which should be her initial tack?
(b) What time should she tack to reach the buoy?
(c) What time will she arrive at the buoy?

5.7 At 1135 a boat has just entered the traffic separation scheme and is 1 mile east of CS3 buoy south east of Dungeness. The wind is NE 4, leeway is 5°.
Tidal stream:
    1135 – 1235   205°T 1.4 knots

(a) What is the magnetic course to a position 1 mile east of Bullock Bank north cardinal buoy if she maintains a speed of 5.5 knots?
(b) What time will she arrive at this position?
(c) Is this course satisfactory?

5.8 At 2250 a boat is in a position 1 mile south of Vergoyer SW cardinal buoy. The boat's speed is 5.5 knots. The wind is NW 4, leeway is negligible.
Tidal streams:
    2220 – 2320   247°T 0.9 knots
    2320 – 0020   235°T 1.2 knots

0020 – 0120   224°T 1.3 knots

(a) What is the magnetic course to a position 50°20′.4N, 1°11′.4E?

(b) What is the ETA at this position?

5.9 At 1650 a boat takes departure from a RW buoy in Rye Bay for a position 50°52′.4N, 0°59′.8E. The boat's speed is 5 knots. The wind is westerly, leeway is negligible. HW Dover is at 1719 and it is a neap tide. Use tidal diamond E.

(a) What is the magnetic course to this position?

(b) At what time will the boat reach this position?

(c) From this position what is the magnetic course to East Road buoy, (north east of Dungeness)? The boat's speed is 6 knots. Leeway is 5°.

(d) What is the ETA at East Road buoy?

5.10 At 0717 a boat has just left the traffic separation scheme and is 1 mile south of Vergoyer E cardinal buoy. She wishes to reach a buoy (Fl G 4s) to the west of Boulogne harbour. The wind is N 4, leeway is 10°. She can sail closehauled 45° off the wind on either tack, maintaining a speed of 4.8 knots. HW Dover is at 1117, and it is a spring tide. Use tidal diamond L.

What is the ETA at the buoy?

CHAPTER SIX

Fixing Position

6.1 At 0740, log 42.1, a boat in DR position 50°44′.0N, 1°30′.2E is approaching Boulogne Harbour when she takes the following bearings:

S. Cardinal buoy	062°M
Green buoy	144°M
ZC1 buoy	283°M

(a) How far is the boat off Boulogne south breakwater lighthouse?

(b) What is her bearing from the breakwater lighthouse?

6.2 At 2230, log 35.1, a boat in DR position 50°53′.6N, 1°28′.1E observes a light, Fl 5s, in transit with another light, Oc (3) Y 12s, bearing 123°C. A sounding taken at this time shows 41 m.

(a) What is the boat's position?

(b) What is the deviation on the compass? (Do not use the standard deviation table.)

6.3 At 1940 a boat in DR position 50°34′.7N, 1°29′.0E, steering a course of 169°C, observes Pt du Touquet light (Fl (2) 10s) bearing 123°C. At 2040 the same light bears 076°C. The boat's speed is 3.5 knots. Leeway is negligible and the tidal stream is slack. Deviation is 5°E.

(a) What is the boat's position at 2040?

(b) What is the reliability of the fix?

6.4 At 2040 a boat in DR position 50°39′.7N, 1°00′.6E observes a light Fl (3) 15s on a bearing of 087°M dipping on the horizon. The height of the tide is 4.6 m. Height of eye is 3.0 m. Plot the 2040 position.

6.5   At 1835, just after sunset, a boat in DR position 50°45′.3N, 1°07′.2E takes the following bearings:

  Light VQ   033°M
  Dungeness RC 331°M

(a) Plot the fix.
(b) Comment upon the reliability of both bearings.

6.6   At 1241 a boat approximately 1.2 miles south of Dungeness new lighthouse takes a vertical sextant angle between the focal plane of the light and sea level which gives 1°17′.3; at the same time the lighthouse bears 352°M. The height of tide at this time is 6 m. Plot the 1241 position.

6.7   At 1015 a boat in DR position 50°44′.8N, 0°20′.3E on a course of 066°C takes the following bearings using the steering compass:

  Chimney approx. 1 mile SW of Langney Point 349°C
  Wish tower   298°C
  Radio tower at Beachy Head   253°C
  No deviation table is available.

(a) What is the 1015 position?
(b) What is the deviation of the steering compass for the boat's present heading?

6.8   At 1620, log 8.4, a boat in DR position 50°42′.1N, 0°12′.2E on a course of 100°C, observes Beachy Head lighthouse bearing 031°M. The wind is N 4, leeway is 5°. The compass deviation is 2°W. At 1750, log 14.7, Royal Sovereign lighthouse bears 316°M. Tidal stream throughout:

  068°T   1.8 knots

Plot the boat's position at 1750.

6.9   At 2240, log 18.1, a boat fixes her position by taking the following bearings:

  Light Fl R 20s   018°M
  Cap Gris-Nez RC 142°M
  Dungeness RC   255°M

She then steers a course of 310°C until 2340, log 23.1, when the following bearings are obtained:

  Light Fl 7.5s   014°M
  Light Fl R 20s   137°M

Light Fl (2) 10s   278°M
Leeway is negligible. Deviation is 5°E.

(a) What is the 2340 position?
(b) What has been the set and drift for the last hour assuming the log is correct?

6.10 At 2010 a boat on passage to Calais in EP position 50°46′.6N, 1°31′.8E, takes the following bearings using the steering compass:

Cap Gris-Nez lighthouse   018°C
East cardinal buoy   141°C
Boulogne north breakwater lighthouse   157°C

Her course is 008°C. The deviation table which has not been checked for some time shows a deviation of 2°W on this heading. At the time of the fix, Cap Gris-Nez lighthouse is observed to be in transit with Bassure de Baas north cardinal buoy.

(a) What is the 2010 position?
(b) How much reliance should you place on this position?

CHAPTER SEVEN

Tides

Use extracts on pages 107 onwards.

7.1 At what time during the morning of 3 August will the height of tide at Dover fall to 3.0 m?

7.2 What is the height of tide at Dover at 1230 on 15 August?

7.3 What is the height of tide at Ramsgate at 0400 on 10 August?

7.4 What is the latest time in the afternoon of 9 August that a boat, with a draught of 1.0 m, can enter Ramsgate harbour with a clearance of 0.5 m?

7.5 What will be the depth of water alongside the south quay at Folkestone at 0618 on 19 August?

7.6 At 1653 on 17 August a boat anchors just west of Folkestone in a depth of 5.0 m. What will be the greatest and least depth in the anchorage over the next 24 hours?

7.7 At 0830 on 3 August, a boat in DR position 51°03′.9N, 1°13′.5E is heading for Dover. From this position she runs a line of soundings. What is the reduction to apply to these soundings?

7.8 At 0330 on 12 August, a boat in DR position 51°06′.0N, 2°02′.0E, observes a light bearing 280°M, with characteristics Fl (3) 20s, dipping on the horizon. Height of eye is 3.0m. What is the distance off the light?

7.9 What is the height of tide at Yarmouth at 0030 on 24 April?

7.10 What is the depth of water over the bar at Lymington entrance on 19 April at 1231?

7.11 At 0634 on 17 April a boat goes aground at Lymington entrance. When will she refloat?

CHAPTER EIGHT

Pilotage and Passage Planning

The following questions are based on the pilotage information given in Extract 19 and instruction chart 5055. The characteristics of several of the lights and buoys in Calais approaches have been changed since the pilotage information in Extract 19 was compiled, hence there are some discrepancies between that information and the 1987 edition of chart 5055. Where appropriate use the information in Extract 19 to answer the questions.

8.1 What Admiralty charts would be required to cover the approach and entrance to Calais?

8.2 What navigational hazards might be expected when approaching Calais from the east?

8.3 (a) On 25 August between what times, during the day, will the tidal stream be fair when approaching Calais from the west?
(b) When will the tidal stream run at its greatest rate?
(c) What will that rate be?

8.4 Where can the times of HW Calais be found?

8.5 If HW Calais is at 1620, at what times could a motor boat enter the yacht harbour at Calais?

8.6 Before leaving a UK port, is it possible to enquire about the availability of a berth in the yacht harbour at Calais?

8.7 Approaching Calais, when 0.5 miles off, how is it possible to check the likelihood of a ferry leaving the harbour?

8.8 A motor boat expects to arrive at Calais near the time of local LW. Where can she wait until the yacht harbour is accessible?

8.9 (a) Approaching the entrance to the Calais harbour at night, list the lights that are likely to be visible.

(b) What transits and clearing bearings could be used to ensure that adequate allowance is made for any tidal stream setting across the entrance?

8.10 On 22 August a boat wishes to make a day passage from Boulogne to Calais. Her cruising speed is 6 knots. The weather forecast at 0555 is: wind SW 4–5; visibility moderate with fog patches. Make an outline passage plan.

8.11 When rounding Cap Gris-Nez, it is desired to keep approximately 1 mile offshore. How might this be achieved?

8.12 On a calm day a yacht is motoring from Calais to Boulogne. Cap Blanc-Nez is abeam when the visibility drops to about 0.5 miles. What navigational action should be taken?

CHAPTER NINE

Coastal Passage Making

Test Paper A

On 6 August a bilge-keel motor sailing boat, draught 1.0 m, is making a passage from Newhaven to Rye. She carries an echo sounder, radio direction finder and an accurate log. She does not carry a VHF radio. The weather forecast is wind variable 3 or less, visibility good. Her normal cruising speed is 4.5 knots.

A.1 At 0500 the following bearings are taken:

      Royal Sovereign light   018°M
      Beachy Head lighthouse   288°M

At this time a sounding of 13 m is taken.

    (a) Plot the position.
    (b) Comment on the accuracy of the fix.

A.2 From the 0500 position:

    (a) What is the course to steer for the RW buoy in Rye Bay?
    (b) What is the ETA at the buoy?

The boat's speed is 4.5 knots. Leeway is negligible. Use tidal diamond C.

A.3 Using the course calculated in question A.2, plot the EP for 0720. The boat's speed is 4.5 knots. Leeway is negligible.

A.4 The same course and speed are maintained until 0800 when the following bearings are taken:

      Camber Church   010°M
      Dungeness RC   085°M

Plot the fix.

A.5 The RW buoy is reached at 0807.

(a) What is the course to steer from the buoy to Rye harbour entrance?

(b) What will be the ETA at the entrance?

The boat's speed is 4 knots. The tidal stream and leeway are negligible.

A.6 (a) Is it likely that the boat will be able to enter Rye harbour at the ETA obtained in question A.5?

(b) What precautions should she take if she enters?

(c) What are the alternatives to entering?

A.7 How would your answer to question A.6 be affected if the boat carried a VHF radio?

Test Paper B

B.1 A boat is heading for Calais on a course of 351°C. Her 0930 DR position is 50°19′.3N, 1°30′.1E. At 0930, log 84.2, she obtains the following bearing:

       Pt du Haut Banc lighthouse   042°M

At the same time a west cardinal buoy is just visible abeam to starboard, the log has been behaving erratically for the last hour and is probably unreliable. At 1030, log 87.2, the same lighthouse bears 107°M and a sounding taken at this time indicates 13 m. The wind is E 4, leeway is 10°. HW Dover is 1300 and it is a spring tide. Use tidal diamond K.

(a) Plot the EP at 1030.

(b) Plot the fix at 1030.

(c) Comment on the fix obtained.

B.2 The 1030 position is 50°24′.7N, 1°28′.2E. At 1030 the boat alters course to 014°C. Speed is estimated at 4 knots. The wind is E 4, leeway is 10°. At 1130, log 90.2, the following bearings are taken:

       Pt du Touquet lighthouse   071°M

       Stella Plage Church   112°M

(a) Plot the EP at 1130.

(b) Plot the fix at 1130.

Use tidal diamond K.

B.3 It is evident that the log is unreliable and an estimated speed of 4 knots is used for further plotting. The boat continues on the same course. The wind is E 4, leeway is 10°. Between 1130 and 1230 the following soundings are obtained:

1139	9.8 m
1150	8.9 m
1158	12.4 m
1221	16.0 m
1230	15.0 m

    (a) Plot the EP at 1230.
    (b) Do the soundings confirm the ground track?
Use tidal diamond K.

B.4 (a) What is the course to steer from the 1230 EP to the green buoy west of Boulogne entrance, assuming the speed is 4 knots, wind E 4 and leeway is 10°?
    (b) What is the ETA at the buoy?
Use tidal diamond L.

B.5 The boat passes the buoy at 1358 and at 1404 takes the following bearings:

South cardinal buoy	040°M
Boulogne south breakwater lighthouse	093°M
Green buoy	141°M

Plot the fix at 1404.

B.6 The wind has been decreasing for the last hour and from the 1404 position the boat motors on a course of 029°C making a speed of 6 knots. Leeway is negligible. At 1410, a south cardinal buoy is passed close to starboard. At 1431, Bassure de Baas north cardinal buoy is in transit with Cap Gris-Nez lighthouse and Wimereux Church bears 134°M. A sounding taken at this time indicates 10.0 m.

    What has been the set and drift of the tidal stream between 1404 and 1431?

B.7 Visibility is beginning to deteriorate and it is decided to return to Boulogne rather than continue the passage against a foul tidal stream with a useless log. The boat can maintain 6 knots under power. Leeway is negligible.
    (a) What is the course to steer for Boulogne south breakwater?

(b) What will be the ETA at the breakwater?
Use tidal diamonds L and N.

Test Paper C

C.1 At 1040, log 31.0, a motor boat is in a position 1 mile south
of Dungeness lighthouse. She wishes to pass through a position 1.2
miles to the south of Beachy Head lighthouse. Her cruising speed
is 4.5 knots. The wind is NW 3. Visibility good. Leeway is
negligible. Use the tidal streams given after question C.5.
(a) What is the course to steer to this position?
(b) What is the ETA at this position?

C.2 The course found in question C.1 is steered. At 1340, log
44.5, Royal Sovereign lighthouse bears 237°M. At 1510, log 51.3,
the same lighthouse bears 190°M.
(a) Plot the EP at 1510.
(b) Plot the fix at 1510.
(c) Is the log correct?

C.3 At 1510 course is altered to 248°C. Plot the EP at 1640 when
the log reading is 58.1.

C.4 At 1640 the following bearings are obtained:
Beachy Head lighthouse   042°M
Newhaven RC   311°M

(a) Plot the 1640 position.
(b) Does this position agree with the EP?

C.5 (a) What is the course to steer from the 1640 position to
Newhaven breakwater lighthouse? The boat's speed is
5.5 knots.
(b) What is the ETA at the breakwater?
Tidal streams:
1040 – 1140   031°T 1.5 knots
1140 – 1240   031°T 1.9 knots
1240 – 1340   248°T 1.4 knots
1340 – 1440   248°T 1.8 knots
1440 – 1540   248°T 1.7 knots
1540 – 1640   263°T 1.8 knots
1640 – 1740   267°T 1.3 knots

Test Paper D

D.1 A motor sailer with a draught of 1 m is berthed in Folkestone Harbour on a berth which dries 1.6 m. What is the latest time during the morning of 31 August that she can leave her berth with a clearance of 0.5 m?

D.2 The proposed passage is from Folkestone to Rye staying overnight at Rye. What is the best time to leave Folkestone? The boat's speed is 6 knots. The weather forecast: NW 5, decreasing to 2 later, visibility good. Use the tidal streams given after question D.7.

D.3 The boat leaves her berth at 0735 and at 0800 passes very close to a north cardinal buoy (Sandgate). The boat's speed is 6 knots. Leeway due to the north westerly wind is 5°.

    (a) What is the course to steer from this buoy to a position 1 mile south of Dungeness lighthouse?
    (b) What is the ETA at this position?

D.4 At 0901 a red buoy bears 315°M at an estimated distance of 0.7 miles. Does this confirm the boat is on the required track?

D.5 This course is maintained and at 0958 Dungeness lighthouse bears 016°M. A sounding taken at this time gives 20.4 m. Plot the position.

D.6 At 0958 course is altered to 282°C to reach the red and white buoy in Rye Bay. The wind is decreasing and the rest of the passage is made under motor. The speed is 6 knots, leeway is negligible.

    (a) What time will the red and white buoy be abeam?
    (b) Which side of the boat will it be?
    (c) How far off will it be?

D.7 The boat cannot enter Rye harbour due to insufficient depth of water and decides to anchor in Rye Bay. What general considerations should be taken into account before doing so?
Tidal streams:

0800	059°T	0.6 knots
0900	245°T	0.2 knots
1000	211°T	0.8 knots
1100	211°T	1.0 knots

1200   211°T 1.3 knots
1300   211°T 1.2 knots
1400   211°T 0.6 knots
1500   slack
1600   031°T 0.6 knots

Test Paper E

E.1 On 8 August a boat is in Wellington Dock at Dover. She wishes to leave Dover Harbour in time to catch the earliest west going stream that afternoon. At what time should she leave the dock?

E.2 The boat leaves the harbour at 1530. The following is an extract from her log:

Time	Course	Log	Wind	Leeway	Remarks
1600	188°C	0.0	SW 5	10°	1 mile south of Dover west breakwater lighthouse on starboard tack
1640	271°C	3.9	SW 5	10°	Tacked to port tack
1800	188°C	11.9	SW 6	10°	Tacked to starboard tack
1830	271°C	15.3	SW 6	10°	Tacked to port tack
1900	188°C	18.4	SW 5	10°	Tacked to starboard

Use tidal diamond J until 1800 and then use tidal streams as follows:

1830   211°T 1.6 knots
1930   211°T 2.1 knots
2030   211°T 1.8 knots
2130   211°T 0.9 knots

(a) What is the EP at 1800?
(b) What is the EP at 1900?

E.3 The course of 188°C is maintained until 1930, log 21.4. Plot the EP at 1930. Leeway is 10°.

E.4 (a) What is the course to steer from the 1930 EP for the RW buoy in Rye Bay? The boat's speed is 6 knots. The wind is SW 5, leeway is 10°.
(b) What is the ETA at this position?

CHAPTER TEN

Offshore Passage Making

Test Paper F

On 19 August a motor sailing boat is in a harbour at Le Touquet and she must leave shortly before HW because of the depth in the approach channel. She plans to sail to Folkestone. She has an echo sounder and a radio direction finder (RDF) on board.

F.1 She leaves the harbour at 0900 and at 0935, log 3.2, is in position 50°34'.4N, 1°33'.0E steering a course of 347°C. The wind is E 4, leeway is 5°.

   Plot the EP when the log reads 11.5.
Use tidal diamond L.

F.2 At 1105 the following bearings are obtained using an azimuth ring on the steering compass:

South cardinal buoy	039°C
Cap d'Alprech lighthouse	132°C

   Plot the fix.

F.3 From the 1105 position:

   (a) What is the course to steer for ZC2 yellow buoy approximately 3 miles north west of Cap Gris-Nez?
   (b) What is the ETA at the buoy?
The boat's speed is 5.5 knots. The wind is E 4, leeway is 5°. Use tidal diamond N.

F.4 The boat arrives at the buoy at 1225, log 18.5, and alters course to 305°C to cross the traffic separation scheme. Leeway and tidal stream are negligible. At 1305, log 21.8, the following bearings are taken:

Cap Gris-Nez RC	130°M
Dungeness RC	271°M
South Foreland RC	351°M

Plot the fix at 1305.

F.5 The boat continues on the same course. The log is now indicating speed only, which is 5.0 knots. Leeway is negligible. The tidal stream is slack. At 1337 she passes a north cardinal buoy approximately 0.2 miles to starboard. Between 1355 and 1402 the following soundings are taken:

> 1356 – 20 m
> 1359 – 10 m
> 1400 – 5 m
> 1401 – 10 m
> 1402 – 20 m

At 1405, a green buoy is abeam close to starboard. Is the log indicating the correct speed?

F.6 The boat continues on the same course, speed 5.6 knots. Leeway is negligible. At 1505 the following bearings are taken:

> Dungeness RC   235°M
> North cardinal buoy   332°M

Tidal stream:

> 1405 – 1505   233°T 0.6 knots

(a) Plot the EP at 1505.

(b) Plot the fix at 1505.

F.7 From the 1505 position:

(a) What is the course to steer to Folkestone breakwater lighthouse?

(b) What is the ETA at the lighthouse?

The speed is 5.6 knots. The wind is E 4, leeway is 5°.
Tidal stream:

> 1505 – 1605   224°T 0.9 knots

Test Paper G

A sailing boat is on an overnight passage from Dieppe to Rye on 9/10 August. She is fitted with an echo sounder and a radio direction finder. The weather forecast is wind NE 4, veering SE 4–5, becoming southerly later, visibility moderate.

G.1 At 2359 on 9 August, log 134.2, the boat is in position 50°17′.9N, 1°09′.4E. The wind is NE 4 and she is making a speed of about 4.5 knots and 10° leeway. From this position:

    (a) What is the course to steer for the RW buoy in Rye Bay?

    (b) What is the ETA at the buoy?

(For planning purposes, ignore the traffic separation scheme.) Use tidal diamonds F (for the first 4 hours), D (for the following 3 hours), and E (for the remainder).

G.2 From the 2359 position the boat steers a course of 343°C. Leeway is 10°. At 0200 on 10 August the log reads 143.5. What is the EP at 0200?

Use tidal diamond F.

G.3 At 0200 a radio bearing of 358°M is obtained on a frequency of 310.3 kHz, station callsign DU. A sounding of 23 m is taken at the same time. Do these confirm the EP?

G.4 The boat continues on the same course making 10° of leeway. At 0300, log 148.9, a light (VQ (3)), in transit with a more distant light (Fl (4) 15s), is passed close to port. Plot the EP and the fix at 0300.

Use tidal diamond F.

G.5 The same course is continued but the log has stopped reading at 148.9. The wind veers to SE 5 over the next 2 hours and the average speed between 0300 and 0500 is estimated as 6 knots, leeway is 5°. A sounding at 0500 gives 31 m. Plot the EP at 0500.

Use tidal diamond D.

G.6 The wind continues to increase and veers to S 6. Should any action be taken to check the boat's position?

G.7 The course is maintained. Speed is estimated at 6 knots. Leeway is negligible. At 0530 the following radio bearings are obtained on a frequency of 310.3 kHz:

Callsign	Bearing
DU	043°M
RY	257°M

(a) Plot the EP at 0530.
(b) Plot the fix at 0530.

Use tidal diamond D.

The log is now working again. The reading at 0530 is 148.9.

G.8 It had been intended to anchor off Rye and enter the harbour about 1 hour before HW.

(a) Is this still advisable?
(b) What are the alternatives?

G.9 At 0530 course is altered to 247°C. Leeway from the southerly wind is 5°. At 0830 the log reads 166.9. Plot the EP for 0830.

Use the tidal streams given in the table after question G.12.

G.10 At 0830 the course is altered to 273°C. The wind is S 6, leeway is 5°. At 0900, log 169.7, the following bearings are taken:

Royal Sovereign lighthouse	073°M
Beachy Head lighthouse	321°M

Plot the 0900 position.

G.11 From the 0900 position:

(a) What is the course to steer for Newhaven breakwater?
(b) What is the ETA at the breakwater?

Assume a speed of 6 knots and 5° of leeway from a southerly wind.

G.12 On approaching Newhaven breakwater, how can it be determined whether a cross channel steamer may be arriving or departing?

Tidal streams:

To be used from 0530

0530 – 0630	232°T 1.2 knots
0630 – 0730	248°T 0.8 knots
0730 – 0830	067°T 0.5 knots
0830 – 0930	077°T 1.9 knots
0930 – 1030	080°T 2.4 knots

$$1030 - 1130 \qquad 075°T \ 1.4 \ knots$$
$$1130 - 1230 \qquad 107°T \ 0.2 \ knots$$

Test Paper H

A sailing boat with a powerful auxiliary engine is making an overnight passage on 12 and 13 August along the French coast from Dieppe to her home port at Folkestone. The boat's draught is 1 m. Her berth at Folkestone dries 1.5 m above chart datum. The weather forecast is wind NE 4, weather fair, visibility moderate to good. She is equipped with an echo sounder and radio direction finder but no VHF radio.

H.1 At 2030, log 51.9, on 12 August in DR position 50°20′.9N, 1°25′.8E, she raises a light with characteristics Fl (3) 15s on a bearing of 016°M. Assume a height of eye of 1.5 m and a height of tide at Boulogne of 1.1 m. Plot the boat's position.

H.2 From the 2030 position the course steered is 359°C on starboard tack making 10° of leeway. At 2230, log 61.1, the following bearings are obtained:

Cap d'Alprech light	032°M
Pt du Touquet Aero RC	069°M
Pt du Haut Banc light	126°M

  (a) Plot the EP at 2230.
  (b) Plot the fix at 2230.
  (c) Does the EP agree with the fix?
Use tidal diamond K.

H.3 At 2230 the boat tacks to a course of 101°C making 10° of leeway. Her speed is 4.5 knots. She has been sailing around 45° off the wind but has an inexperienced helmsman so it is decided that when closehauled a course 50° off the wind represents the actual course sailed on either tack. Leeway on both tacks is 10°.

  (a) At what time should the boat tack to reach ZC1 buoy west of Boulogne?
  (b) What is the ETA at the buoy?

Use tidal diamond K for the first two and a half hours and tidal diamond L for the remainder.

H.4 At 2348 the boat tacks on to a course of 354°C, leeway is 10°. At 2359, log 67.6, she fixes her position as 50°27′.0N, 1°30′.8E. Plot the position.

H.5 The boat continues on the same course. At 0030, log 70.1, Pt du Touquet light bears 071°M. The same course is maintained and at 0130, log 74.5, the same light bears 142°M. Plot the fix at 0130. Use tidal diamond K.

H.6 The same course is steered until 0230, log 78.6, when the following bearings are obtained:

       Cap d'Alprech light   069°M
       Pt du Touquet light   158°M

(a) Plot the fix.
(b) What has been the set and drift of the tidal stream during the last hour?

H.7 The boat carries on on the same course until 0255, log 80.6, when course is altered to 302°C to cross the traffic separation scheme. Leeway is 5°. Plot the EP at 0400 log 85.8.
Use tidal diamond L.

H.8 The same course is maintained, speed 5 knots. Between 0400 and 0500 the following soundings are obtained:

       0433   20 m
       0438   3 m
       0442   20 m

Do these soundings agree with the 0500 EP at which time the log reads 90.8?
Use tidal diamond G.

H.9 If the boat maintains her course and speed, when may she expect to clear the traffic separation scheme?
The tidal stream for the next hour and a quarter is:

       048°T 1.3 knots

H.10 At 0625, log 97.6, she takes the following bearings:

       South Foreland RC   033°M
       Cap Gris-Nez RC   106°M
       Dungeness RC   265°M

     A sounding at 0625 gave 20 m

Plot the fix.

H.11 The wind has dropped and it is decided to motor making a speed of 6 knots. Leeway is negligible.

> (a) What is the course to steer to the entrance to Folkestone harbour?

> (b) What will be the ETA at the harbour entrance?

Tidal streams:

> 0625 – 0725   011°T 1.2 knots
> 0725 – 0825   Slack

H.12 (a) Can she reach her berth at the ETA found in question H.11 with a clearance of 0.5 m?

> (b) If not what is the earliest time after arrival that she can proceed to her berth?

Test Paper I

This exercise concerns the passage of a sailing boat with an auxiliary engine from Le Touquet (south of Boulogne) to Folkestone. The passage is to commence during the morning of 4 August and the destination must be reached by sunset on 6 August. The boat has a draught of 1.0 m and wishes to reach her berth, which dries 1.6 m above chart datum, with a least clearance of 0.5 m. She can leave the marina at Le Touquet from 1 hour before high water to high water. On 4 August high water Le Touquet is 0704 BST. The shipping forecast for sea area Dover is, wind SW 5, weather fair, visibility good. The boat is equipped with a radio direction finder, an echo sounder and a VHF radio.

I.1   Outline a simple passage plan.

I.2   The boat leaves the marina at 0702 and at 0740 is in position 50°35′.7N, 1°32′.6E.

> (a) What is the course to steer to reach a position 270°T Boulogne south breakwater lighthouse 0.8 miles?
> (b) What is the ETA at this point?

The boat's speed is 5 knots. The wind is SW 5, leeway is 5°. Use tidal diamond L.

I.3 The boat arrives off Boulogne harbour entrance at 0910 when the decision is made to enter the harbour and stay overnight.

(a) What are the harbour entry signals permitting entry into the outer harbour?
(b) Where will they be displayed?

I.4 Assuming a passage speed of 5 knots and wind SW 4, at what time on the morning of 5 August should the boat leave Boulogne to complete her passage to Folkestone?

I.5 The boat leaves Boulogne harbour at 0650 on 5 August and at 0720 is 0.7 miles north of the north breakwater lighthouse. The log is set to zero. Plot the boat's position.

I.6 Assuming a speed of 5 knots, wind SW 4 and leeway of 5°:

(a) What is the course to steer from the 0720 position to a position 50°48′.7N, 1°28′.9E?
(b) What is the ETA at this position?

Use tidal diamond N.

I.7 At 0820, log 5.0, the following bearings are obtained:

Cap Gris-Nez lighthouse   058°M
Cap d'Alprech RC   160°M

Plot the fix.

I.8 From the 0820 position a course of 302°C is set. The wind is SW 6 and leeway is estimated at 5°. At 0920 the log reading is 9.9. Plot the EP.

Use tidal diamond N.

I.9 The boat continues on the same course with leeway constant at 5°. At 0950, log 12.4, the following bearings are obtained:

Cap Gris-Nez RC   118°M
Cap d'Alprech RC   154°M
Dungeness RC   268°M

(a) Plot the fix.
(b) What has been the set and drift of the tidal stream between 0820 and 0950?

I.10 The boat continues on the same course making a speed of 5 knots. At what time will she clear the traffic separation scheme?

Use tidal diamond I.

I.11 At 1050, log 17.4, the following bearings are taken:

>   Cap Gris-Nez RC   127°M
>   Dungeness RC   248°M

(a) Plot the fix.

(b) Comment on the fix.

I.12 From the 1050 position:

(a) What is the course to steer for the harbour entrance at Folkestone?

(b) What is the ETA at the harbour entrance?

The boat's speed is 5 knots, wind SW 4, leeway is 5°.
Use tidal diamond J.

Test Paper J

A bilge keeled motor sailer, draught 1.0 m, is planning to make a passage from Calais to Rye on 13 August. The boat can cruise under power at 7 knots. She is fitted with an echo sounder, radio direction finder and VHF radio. The weather forecast is wind variable 3 or less, visibility moderate to poor. The boat makes no leeway under power in light winds.

J.1 The decision is made to leave Calais as soon as the tidal stream is fair and to make the complete passage under power at the normal cruising speed. Assuming that the traffic separation scheme is crossed at right angles from ZC2 buoy (north west of Cap Gris-Nez), what should be the approximate ETA in the vicinity of the RW landfall buoy in Rye Bay?

J.2 The boat leaves Calais at 0840 and at 0911 is in position 50°58'.4N, 1°44'.0E.

(a) What is the course to steer to position 50°55'.8N, 1°37'.5E?

(b) What will be the ETA at this position?

The boat's speed is 7 knots. Use tidal diamond S.

J.3 This position is reached at 0944 and a course of 315°C is set.

At what time should the boat be clear of the traffic separation scheme? The boat's speed is 7 knots. Use tidal diamond M.

J.4   At 1130 the engine overheats and is stopped. What action should now be taken? There is no wind and visibility is about 4 miles.

J.5   At 1135 the following bearings are taken:

Varne light vessel	074°M
Cap Gris-Nez RC	136°M

Plot the fix.

J.6   At 1214 the engine problem has been resolved and the engine restarted. Since 1135 the estimated tidal set and drift is 227°T 1.0 mile. What action should be taken now?

J.7   At 1214 the course of 315°C is resumed at a speed of 7 knots. At 1244 the following bearings are taken:

Cap Gris-Nez RC	131°M
Dungeness RC	239°M

(a) Plot the EP at 1244.
(b) Plot the fix at 1244.

From 1214 use the tidal streams tabulated after question J.9.

J.8   (a) From this position, what is the course to steer for a position 180°T Dungeness lighthouse, 1 mile, at a speed of 7 knots?
(b) What is the ETA at this position?

J.9   From 1244 visibility has been deteriorating and at 1400 is estimated at 1 mile. At 1414 the following bearings are obtained:

South Foreland RC	049°M
Cap Gris-Nez RC	103°M
Dungeness RC	253°M

(a) What is the position?
(b) What action should be taken?

Tidal streams:

1215 – 1315	245°T 1.4 knots
1315 – 1415	031°T 0.5 knots
1415 – 1515	031°T 1.2 knots

Answers to Questions

CHAPTER ONE

1.1   (a) **50°54′.0N, 0°48′.1E.**
      (b) **50°43′.4N, 0°26′.1E.**
      (c) **51°01′.2N, 1°24′.0E.**
      (d) **50°53′.5N, 1°31′.0E.**
1.2   (a) **Bassurelle light buoy.**
      (b) **Vergoyer SW cardinal buoy.**
      (c) **Vergoyer E cardinal buoy.**
      (d) **ZC1 buoy.**
1.3   **346°T.**
1.4   **50°41′.8N, 1°29′.0E.**
1.5   **50°43′.9N, 1°31′.2E.**

CHAPTER TWO

2.1   (a) **A preferred channel buoy indicating that the preferred channel is to starboard.**
      (b) **An isolated danger buoy which is moored over the danger.**
      (c) **A starboard hand buoy indicating the starboard side of the channel.**
2.2   (a) **A west cardinal buoy.**
      (b) **It indicates that there is a danger to the east of the buoy.**
2.3   **It exhibits a group of three flashes every 15s. The focal plane of the light is 62 m above MHWS. The nominal range is 23 miles.**
2.4   The light structure is Cap Gris-Nez lighthouse.
      (a) **A white tower.**
      (b) **From 232°T to 005°T.**
2.5   (a) **A special mark.**
      (b) **It is used as a traffic separation mark where conventional channel marking could cause confusion.**

2.6 (a) **The fog signal is a single blast on a horn every 30s.**
(b) **No.** Royal Sovereign lighthouse emits a fog signal consisting of two blasts on a diaphone every 30s.

CHAPTER THREE

3.1 (a) **175°T.**
(b) **198°T.**
(c) **091°T.**
(d) **007°T.**
3.2 (a) **160°M.**
(b) **315°M.**
(c) **043°M.**
(d) **296°M.**
3.3 (a) **329°T.**
(b) **235°T.**
(c) **056°T.**
(d) **151°T.**
3.4 (a) **348°T.**
(b) **255°T.**
(c) **124°T.**
3.5

	Compass error	*Deviation*
(a)	**5°W**	**12°W**
(b)	**6°E**	**1°W**
(c)	**12°W**	**19°W**
(d)	**8°E**	**1°E**

CHAPTER FOUR

4.1 **50°28′.8N, 1°25′.1E.**
4.2 **50°19′.5N, 1°27′.0E.**
4.3 (a) **50°32′.9N, 1°31′.4E.**
(b) **121°T.**
(c) **096°T.**
(d) **7.3** miles.
(e) **3.7** knots.
4.4 **50°30′.0N, 1°06′.6E.**
Course 330°T
Water track 320°T
Distance run 12 miles
4.5 **50°26′.3N, 1°21′.2E.**
Course 184°T

Water track 189°T
Distance run 8.9 miles

4.6 **50°21′.9N, 1°25′.8E.**
Course 120°T
Water track 110°T
Distance run 2.4 miles
Course 130°T
Water track 120°T
Distance run 6.3 miles

4.7 **50°39′.7N, 1°27′.0E.**
Course 310°T (starboard tack)
Water track 300°T
Distance run 7.5 miles
Course 050°T (port tack)
Water track 060°T
Distance run 2.5 miles

4.8 (a) **50°30′.7N, 1°17′.3E.**
(b) **50°32′.3N, 1°11′.3E.**
Course 020°T (port tack)
Water track 030°T
Distance run 5.1 miles
Course 290°T (starboard tack)
Water track 280°T
Distance run 4.9 miles

4.9 (a) **50°19′.4N, 0°47′.9E.**
(b) **4.8 knots.**
Course 320°T (port tack)
Water track 330°T
Distance run 6.0 miles
Course 230°T (starboard tack)
Water track 220°T
Distance run 9.2 miles

4.10 **50°22′.6N, 1°09′.8E.**
Course 045°T (starboard tack)
Water track 035°T
Distance run 7.5 miles
Course 135°T (port tack)
Water track 145°T
Distance run 7.5 miles
Course 028°T (starboard tack)
Water track 018°T
Distance run 2.5 miles

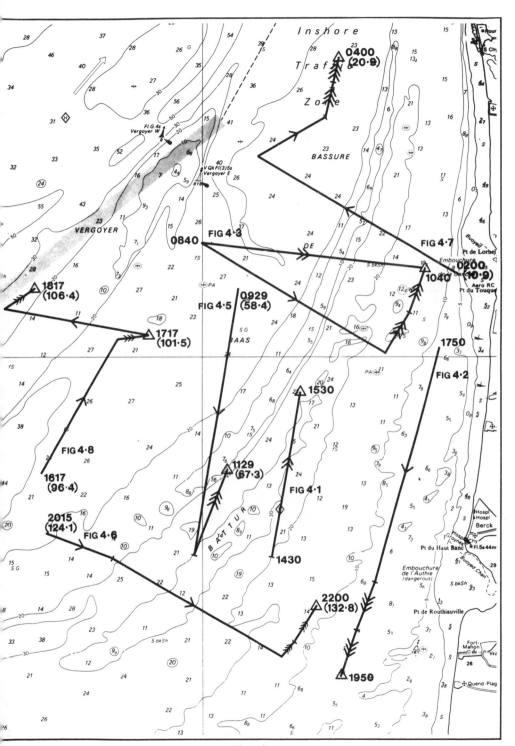

Fig. 4a

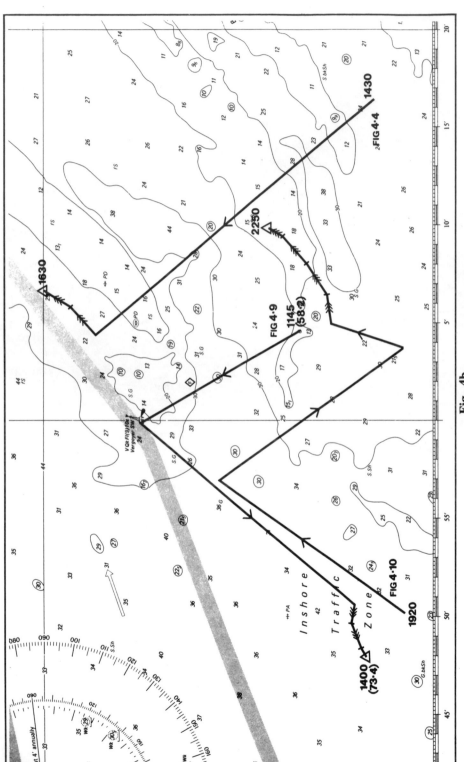

Fig. 4b

CHAPTER FIVE

5.1 (a) Course **024°T**.
(b) ETA **1355**.
Distance to go 5.0 miles
Speed made good 3.0 knots
Time taken 1h 40m
Water track 024°T

5.2 (a) Course **294°M**.
(b) ETA **1834**.
Distance to go 11.8 miles
Speed made good 6.8 knots
Time taken 1h 44m
Water track 290°T

5.3 (a) Course to steer **000°C**.
(b) ETA **1155**.
Distance to go 10.8 miles
Speed made good 6.8 knots
Time taken 1h 35m
Water track 005°T

5.4 (a) Course **049°M**.
(b) ETA **1119**.
(c) Starboard tack
Distance to go 8.9 miles
Speed made good 8.3 knots
Time taken 1h 04m
Water track 035°T

5.5 (a) Course on starboard tack **310°T**
Course on port tack **050°T**
(b) **Either tack.** Whichever tack is chosen initially the ETA will be the same but the time to tack will differ.
(c) Starboard tack **0710**, port tack **0823**.
(d) ETA **0903**.

	Starboard tack	Port tack
Distance to go	3.0 miles	8.5 miles
Speed made good	4.5 knots	4.5 knots
Time taken	0h 40m	1h 53m
Water track	300°T	060°T
Ground track	300°T	060°T

The initial ground track has been plotted from the 0630 position. The reciprocal of the second ground track has been plotted from the buoy. The position to tack is the point at which the two ground tracks intersect.

5.6   (a) Whichever tack the boat starts on she will arrive at the buoy at
       the same time. Preference should be given to **starboard tack**
       initially as, should the boat overstand the buoy on port tack,
       she will have to struggle back against the tidal stream.
       (b) Tack at **0630.**
       (c) ETA **0647.**

	Starboard tack	Port tack
Distance to go	4.5 miles	1.7 miles
Speed made good	1.8 knots	6.0 knots
Time taken	2h 30m	0h 17m
Course	270°T	000°T
Water track	260°T	010°T
Ground track	292°T	029°T

Both water tracks have been laid off from the 0400 position
and the tidal stream applied to find the ground tracks. The
reciprocal of the port tack ground track has been plotted from
the buoy. The position to tack is the point at which the two
ground tracks intersect.

5.7   (a) Course **133°M.**
       (b) ETA **1235.**
       Distance to go 6.1 miles
       Speed made good 6.1 knots
       Time taken 1h 00m
       Water track 134°T
       (c) **Yes:** the course steered crosses the traffic separation scheme at
       right angles.

5.8   (a) Course **119°M.**
       (b) ETA **0039.**
       Distance to go 9.1 miles
       Speed made good 5.0 knots
       Time taken 1h 49m
       Water track 115°T
       Half an hour from each of the tidal streams tabulated has been
       plotted:
              247°T   0.5 miles
              235°T   1.2 miles
              224°T   0.7 miles

5.9   (a) Course **117°M.**
       (b) ETA **1818.**
       Distance to go 7.6 miles
       Speed made good 5.2 knots
       Time taken 1h 28m
       Water track 113°T
       Tidal stream:

031°T 0.8 knots, 031°T 1.1 knots (use half)

This gives 031°T 1.4 miles.
(c) Course **020°M.**
(d) ETA **1912.**
Distance to go 6.4 miles
Speed made good 7.1 knots
Time taken 0h 54m
Water track 021°T
Tidal stream:

031°T 1.1 knot (use half), 031°T 1.0 knot (use half)

This gives 031°T 1.1 miles.

5.10 ETA **1055.**
The boat will have to tack twice to reach the buoy unless she re-enters the traffic separation scheme, which is undesirable. The first tack should be a port tack. A convenient point to tack to starboard would be after 2 hours when the tidal stream is just becoming favourable.

The vector diagram for starboard tack has been plotted from the 0717 position to give a ground track which is then transferred to the 0917 position.

It is estimated that the boat will again tack to port in approximately 1 hour at 1017 and a vector has been plotted for 1 hour from 1017 from the green buoy to find the new port tack ground track. The reciprocal of the ground track thus found has been extended from the green buoy to the point where it intersects the starboard tack ground track. This point will be the actual position to tack.
First two tacks:

	Port tack	*Starboard tack*
Distance to go	8.8 miles	5.0 miles
Speed made good	4.4 knots	4.8 knots
Time taken	2h 00m	1h 02m
Course	045°T	315°T
Water track	055°T	305°T
Ground track	068°T	313°T

Tidal streams:

	Tabulated	*Plotted*
Port tack		
IIW – 4	194°T 2.1 knots (use half)	194°T 1.1 miles
HW – 3	174°T 1.2 knots	174°T 1.2 miles
HW – 2	076°T 0.5 knots (use half)	076°T 0.3 miles

Starboard tack

HW – 2	076°T 0.5 knots (use half)	076°T 0.3 miles
HW – 1	019°T 1.0 knot (use half)	019°T 0.5 miles

Third tack:

Port tack

Distance to go 3.5 miles
Speed made good 5.9 knots
Time taken 0h 36m
Course 045°T
Water track 055°T
Ground track 046°T

Tidal streams:

	Tabulated	*Plotted*
HW – 1	019°T 1.0 knot (use half)	019°T 0.5 miles
HW	007°T 1.7 knots (use half)	007°T 0.9 miles

Tack at 0917 and 1019.

Figs. 5.1, 5.2 and 5.3

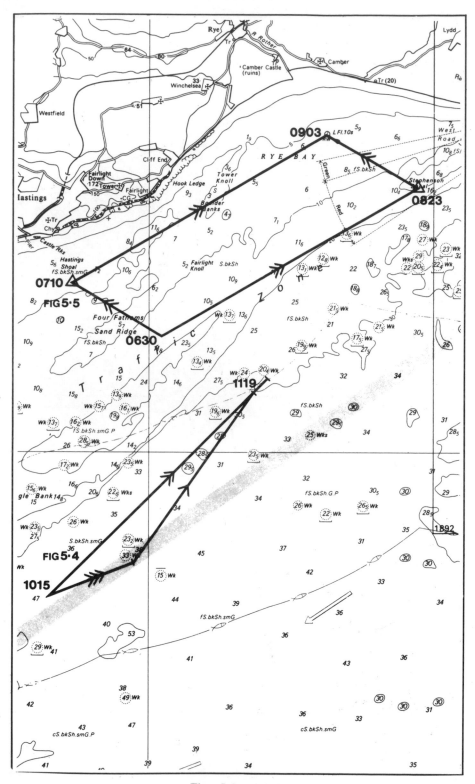

Figs. 5.4 and 5.5

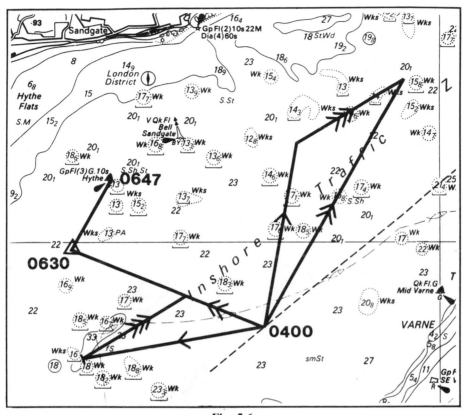

Fig. 5.6

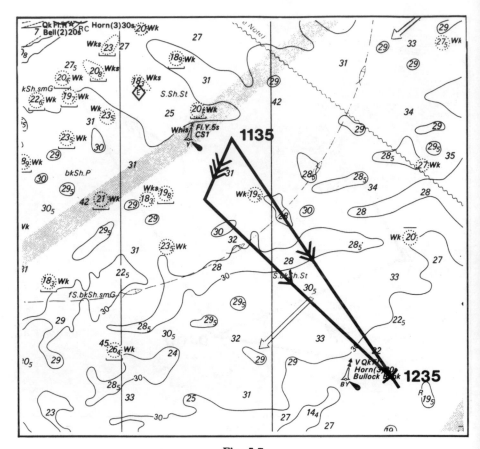

Fig. 5.7

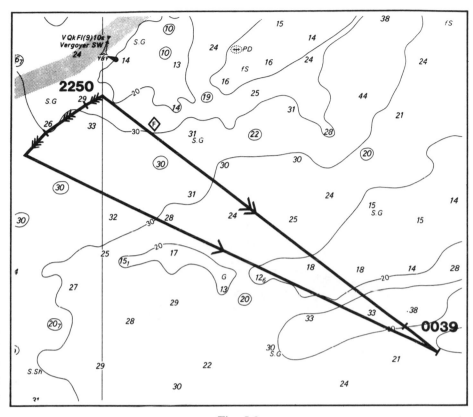

Fig. 5.8

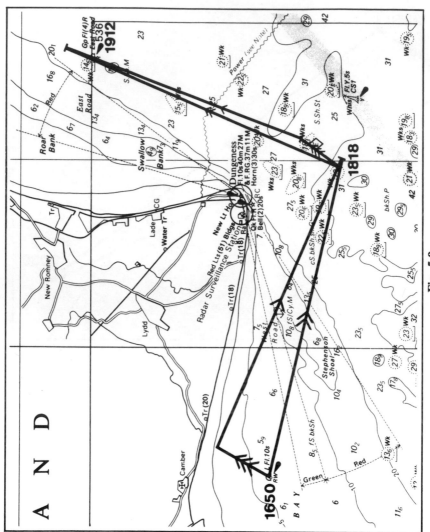

Fig. 5.9

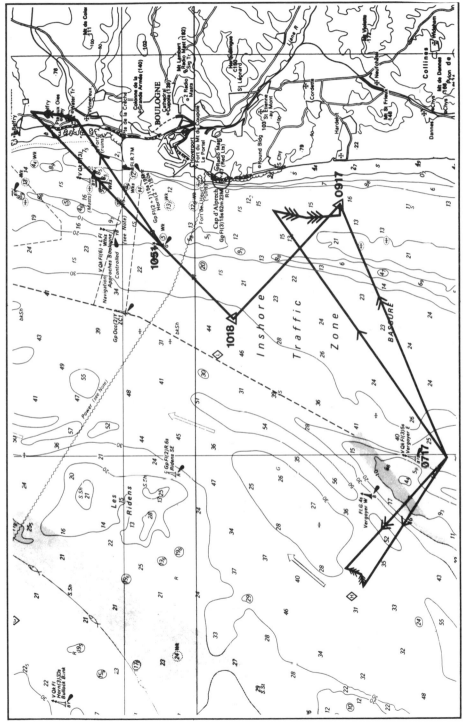

Fig. 5.10

CHAPTER SIX

6.1   (a) **2.8 miles.**
       (b) **271°T.**
       Position 50°44′.4N, 1°29′.8E.
       The cocked hat is normal, three such bearings will rarely intersect
       perfectly. The position within the cocked hat should be assumed to
       be that nearest any imminent danger on the boat's present course.
       If the cocked hat is large the bearings should be retaken.

6.2   (a) The boat is **on the transit line** in the area of the 40 m charted
           depth.
       (b) **1°E.**

6.3   **50°30′.6N, 1°30′.3E.**
       Course 170°T
       Water track 170°T
       The accuracy of a transferred position line depends upon the
       correct assessment of distance run, tidal stream, leeway and course
       steered. It cannot give a position as accurate as one obtained from
       cross bearings. In this case the fix obtained would be acceptable.

6.4   **50°39′.5N, 1°01′.4E.**
       The light is Cap d'Alprech bearing 083°T.

	Metres
Height of light above MHWS	62.0+
MHWS (Boulogne)	8.9
Height of light above CD	70.9−
Height of tide	4.6
Height of light above sea level	66.3

       Distance off   64 = 20.25 miles
                      67 = 20.75 miles
       After interpolation 20.63 miles (20.6 miles)

6.5   (a) **50°46′.4N, 1°07′.2E.**
       (b) The light is Bullock Bank north cardinal buoy. The boat is
           somewhere along the position line from the buoy. Bearings
           using radio direction finding equipment cannot be relied upon
           around dawn or dusk due to the 'sky wave effect'. Bearings
           taken from buoys are less reliable than bearings of shore lights
           or marks at equivalent distances.

6.6 **50°53'.8N, 0°59'.1E**

	Metres
Height of light above MHWS	40+
MHWS	8
Height of light above CD	48−
Height of tide	6
Height of light above sea level	42

Height of light	*Sextant angle*	*Distance off*
40 m	1°14'.0	1.0 mile
43 m	1°19'.0	1.0 mile

Distance off the light is 1.0 mile.

6.7 (a) **50°45'.0N, 0°19'.0E.**

(b) Deviation **6°E**.

Two position circles are plotted: one for the boat, the chimney and Wish tower, and the other for the boat, the radio tower and Wish tower. The boat's position is the point of intersection of these two circles. To construct a position circle, a base line is plotted joining the two landmarks and a line plotted from each landmark at an angle to the base line. This angle is the complement of the angular differences between the bearings. These two lines intersect at the centre of the position circle.

		Angular difference	*90° − Angle difference*
Chimney	349°C	51°	39°
Wish tower	298°C	45°	45°
Radio tower	253°C		

This is an accurate fix, particularly if plotted on a larger scale chart. Deviation is found by comparison of any one of the compass bearings with the true bearing measured from the chart making allowance for variation. This problem can also be done by using a Douglas or Portland protractor or tracing paper. A station pointer would not be practical on this scale of chart.

6.8 **50°42'.7N, 0°27'.4E.**

Distance run 6.3 miles

Course 094°T

Water track 099°T

6.9 (a) **51°04'.2N, 1°19'.0E.**

(b) Set and drift **046°T 1.2 miles.**

Distance run 5 miles

Course 311°T
Water track 311°T
The lights are:

> Varne light vessel
> Folkestone breakwater
> Dover breakwater

6.10 (a) The boat is on the transit between the north cardinal buoy and
Cap Gris-Nez lighthouse. The transit is **019°T**, which gives a
deviation of 5°E and not 2°W as suggested. Using this deviation
of 5°E, the other bearings are **158°T** and **142°T**.

(b) **The size of the cocked hat is unacceptable.** The angle between
Boulogne north breakwater lighthouse and the east cardinal
buoy is too narrow. A better choice would have been Cap
d'Alprech lighthouse instead of Boulogne breakwater lighthouse. As there is no immediate danger, the boat can carry on
to Calais. The echo sounder should be used to check that she is
a safe distance offshore. Frequent fixes should be taken using
the hand bearing compass and use should be made of transits.
On completion of the passage the compass should be swung
and a new deviation table made out.

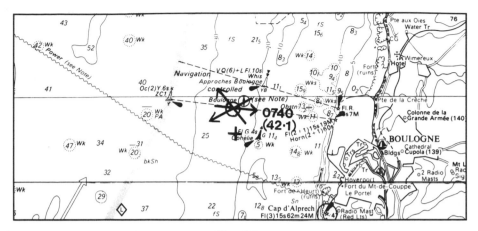

Fig. 6.1

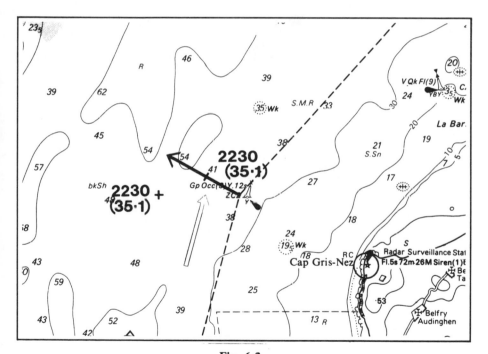

Fig. 6.2

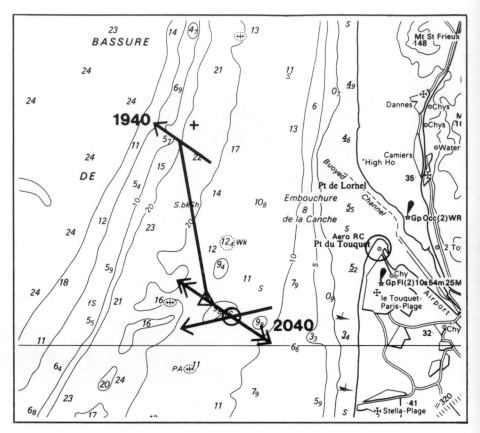

Fig. 6.3

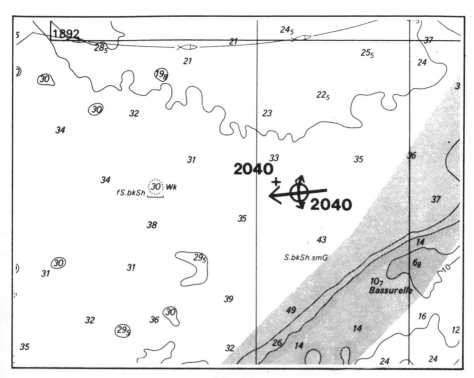

Fig. 6.4

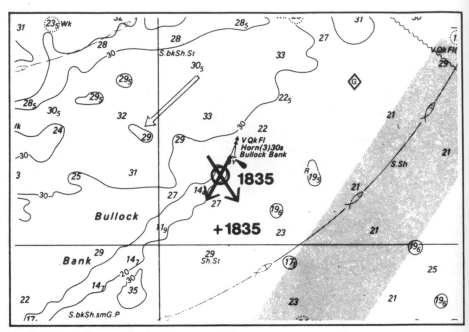

Fig. 6.5

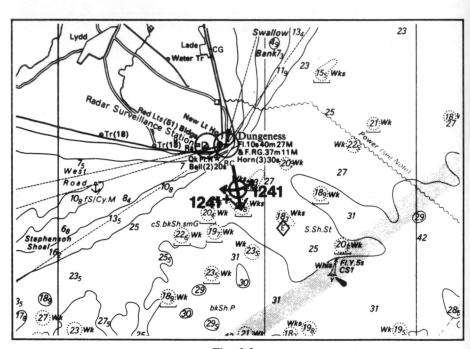

Fig. 6.6

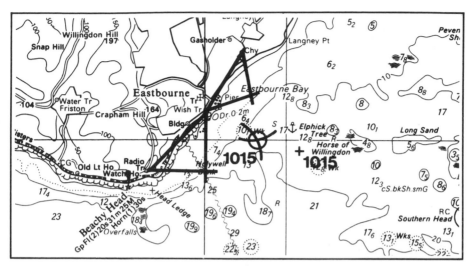

Fig. 6.7

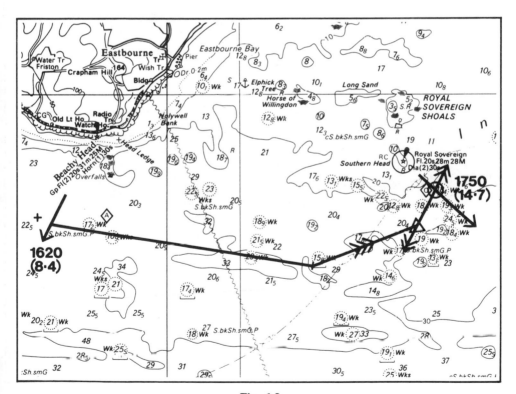

Fig. 6.8

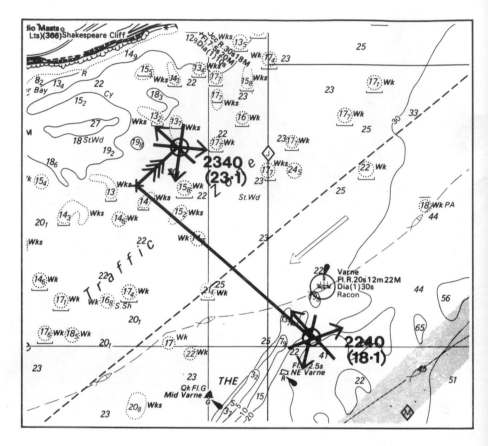

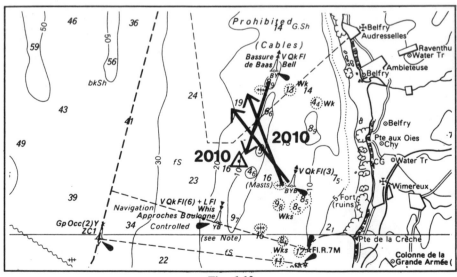

Fig. 6.10

CHAPTER SEVEN

7.1   1044.

		HW		LW
		time	height (m)	height (m)
Dover		0502 GMT	5.4	2.1
		0602 BST		
Range		3.3 m (neap tide)		
Height required		3.0 m		
Interval		4h 42m after HW		
Time		1044		

7.2   1.7 m.

	LW		HW	
Dover	height (m)	time		height (m)
	1.5	1549 GMT		6.1
		1649 BST		
Range	4.6 m (halfway between spring and neap tides)			
Interval	4h 19m before HW			
Height	1.7 m			

7.3   3.1 m.

		HW		LW
		time	height (m)	height (m)
Dover		2333+ GMT	6.7−	0.7−
Differences		0020	1.8	0.4
		———	——	——
		2353	4.9	0.3
		———	——	——
		0053 BST		
Range		6 0 m (spring tide)		
Interval		3h 07m after HW		
Height		3.1 m		

7.4 **1733.**

	HW		LW
	time	height (m)	height (m)
Dover	1115+ GMT	6.7−	0.7−
Differences	0020	1.8	0.4
Ramsgate	1135	4.9	0.3

	1235 BST
Range	6.0 m (spring tide)
Height required	1.4 m (draught + clearance − CD)
Interval	4h 58m after HW
Time	1733

7.5 **2.4 m.**

	LW	HW	
	height (m)	time	height (m)
Dover	2.1	0806− GMT	5.4+
Differences	0.0	0010	0.4
Folkestone	2.1	0756	5.8

	0856 BST
Range	3.3 m (neap tide)
Height	4.0m
Interval	2h 38m before HW
Depth	2.4 m (height − drying height)

7.6 **6.3 m, 2.5 m.**

	LW	HW		LW
	height (m)	time	height (m)	height (m)
Dover	2.2	1805− GMT	5.5+	2.1
Differences	0.0	0005	0.4	0.0
Folkestone	2.2	1800	5.9	2.1

1900 BST

Range	3.3 m (neap tide)
Interval	2h 07m before HW
Height	4.6 m
Depth	5.0 m
Charted depth	0.4 m (depth − height)
Greatest depth	6.3 m (HW height + charted depth)
Least depth	2.5 m (LW height + charted depth at next LW)

7.7 4.6 m.

	HW		LW
	time	height (m)	height (m)
Dover	0502 GMT	5.4	2.1
	0602 BST		
Range	3.3 m (neap tide)		
Interval	2h 28m after HW		
Height	4.6 m		

All readings on the echo sounder should be corrected for the depth of the transducer below sea level. The height of the tide is then subtracted from the resultant figure to give a sounding below CD which is the depth indicated on the chart.

7.8 25.8 miles.

	HW		LW
	time	height (m)	height (m)
Dover	0109 GMT	6.7	0.7
	0209 BST		
Range	6.0 m (spring tide)		
Interval	1h 21m after HW		
Height	6.2 m		

The light is South Foreland.

| Height of light above MHWS (m) | 114.0+ |
| Height of MHWS (m) | 6.7 |

| Height of light above CD (m) | 120.7− |
| Height of tide (m) | 6.2 |

| Height of light above sea level (m) | 114.5 |

From the distance off tables:

Height of light	*Distance off*
107	25.0 miles
122	26.5 miles

After interpolation distance off is 25.8 miles.

The light is just outside its nominal range. However, in good visibility it would still be visible.

7.9 2.3 m.

	HW		LW
	height (m)	time	height (m)
Portsmouth	4.2−	0206− GMT	1.2−
Differences	1.4	0025	0.2
Yarmouth	2.8	0141	1.0
		0241 BST	
Range		3.0 m (half way between spring and neap tides, use the mid curve)	
Interval		2h 11m before LW	
Height		2.3 m	

7.10 3.0 m.

	HW		LW
	height (m)	time	height (m)
Portsmouth	4.0−	0821− GMT	1.0−
Differences	1.5	0020	0.3
Lymington	2.5	0801	0.7
		0901 BST	

Range	3.0 m (halfway between spring and neap tides, use the mid curve).
Interval	3h 30m after LW
Depth over the bar (height + CD)	3.0 m (1.6 + 1.4)

7.11 **0852.**

	HW height (m)	time	LW height (m)
Portsmouth	4.6−	0652− GMT	0.5−
Differences	1.5	0020	0.3
Lymington	3.1	0632	0.2
		0732 BST	

Range	4.1 m (spring tide)
Interval	0h 58m before LW (falling tide)
	1h 20m after LW (rising tide)

The boat will refloat at 0852.

CHAPTER EIGHT

8.1 Admiralty charts **1892** and **1352.**

8.2 **A long sandbank, Ridens de la Rade, extends to the north and the east from Calais harbour.** Approaching from the east, keep about 2 miles offshore until the pierheads are due south, then alter course for the harbour entrance making sure that any eastgoing tidal stream is countered. In strong onshore winds, continue westwards until buoy CA6 can be left to port.

8.3 On 25 August HW Dover is 1303 range 5.4 m. Assume spring tides.

   (a) An ENE tidal stream runs **from HW−1¾ (1118)**, it is slack by **HW + 4½ (1733).**

   (b) At HW **(1303).**

   (c) **2.9 knots.**

8.4 **Admiralty Tide Tables, Reed's Nautical Almanac, Macmillan's Nautical Almanac, local tide tables.**

8.5   The lock gates into the yacht harbour open **from HW −2½ (1350) to HW +1 (1720)**. The swing bridge over the entrance opens at **HW −2½ (1350), HW −1 (1520) and HW +1 (1720)**.

8.6   **Yes:** telephone 010 33 21 34 55 23, which is the yacht club, asking for the enquiry to be passed to the harbourmaster.

8.7   If a VHF radio is carried, **call up on channel 12** (port control) and request clearance to enter the harbour.

8.8   Boats waiting to enter the yacht harbour can either **lie alongside the northern part of the east wall of the Avant Port de l'Ouest** or **pick up a mooring buoy outside the lock gates.**

8.9   (a)

Fl (4) 15s 59m 22M	Main lighthouse in town itself
Fl (2) R 10s 11m 20M	End of east breakwater
Iso G 3s 12m 8M	End of west breakwater
Oc (2) R 6s	CA10 buoy on port hand
FR 14m 10M	Leading light
FG	Inside harbour entrance on starboard hand

(b) The principal transit for the approach is the **Iso G 3s and FR lights in line**. If the tidal stream is running to the west, the **Fl (2) R 10s and the Fl (4) 15s** lights in line is a preferred transit. Clearing bearings may be the Iso G 3s light not less than 094°M and the Fl (2) R 10s light not greater than 184°M.

8.10   On 22 August HW Dover is 1130, range 4.8 m. ENE going tidal stream off Calais from HW −1¾ (0945) to HW +4½ (1600). North going tidal stream off Boulogne from HW −1½ (1000) to HW +4½ (1600). Distance to go is 20 miles. Time taken for passage ignoring the tidal streams, 3h 20m. A sensible time to start is at the beginning of the north going stream at Boulogne, which is 1000. The rate of the tidal stream along the coast is difficult to predict, but it will be favourable throughout the passage.

The passage plan would be:

1000   **Leave Boulogne harbour. Steer 003°M keeping between 1 mile and 2 miles off the coast.**

1020   **East cardinal buoy to port.**

1040   **North cardinal buoy to port.**

1115   **Cap Gris-Nez abeam to starboard. When Cap Gris-Nez bears 138°M, alter course to 038°M. Cap Gris-Nez should be passed about 1 mile distant.**

1200   **West cardinal buoy close to starboard. Alter course to 068°M.**

1225   **CA3 buoy to starboard. Identify the monument on Cap Blanc-Nez. Alter course to 078°M.**

1300   CA4, CA6 and CA10 buoys to port.

1330   ETA entrance to Calais harbour.

8.11 (a) Using a vertical sextant angle (VSA).

(b) Using an echo sounder.

8.12 Take an immediate fix if possible. It would be imprudent to close the coast in the vicinity of Cap Blanc-Nez. CA3 buoy should be approached (either directly if visible, or by setting a course from the last fix or EP position). From CA3 buoy set a course for CA1 buoy. Make a note of the log reading at CA3 buoy and estimate what it should be at CA1 buoy, so that the latter is not missed should the visibility deteriorate further. From CA1 buoy set a course to sail round Cap Gris-Nez using the log to determine the DR position and the echo sounder to remain in a depth of water of at least 10 m. Use the radio direction finder (RDF) to approach Cap Gris-Nez. There is no danger when rounding Cap Gris-Nez if the visibility remains greater than 0.5 miles. Similarly from Cap Gris-Nez, the log, the echo sounder and the RDF can be used to approach Boulogne.

Answers to Test Papers

CHAPTER NINE

Test Paper A

A.1 **50°42′.3N, 0°25′.6E.** The fix obtained by the two position lines is confirmed by the sounding; it may be relied on.

A.2 (a) **044°C.**

(b) **0804.**

The distance to go is 18.4 miles which will take just over 4 hours not allowing for the tidal stream. HW Dover is 0934, range 4.1 m. It is one third of the way from a neap tide to a spring tide.

Tidal streams:

Tidal diamond C

0500 – 0600	068°T	1.3 knots
0600 – 0700	068°T	1.9 knots
0700 – 0800	068°T	1.6 knots

Thus the tidal streams for the first three hours will shorten the passage by the equivalent of 4.8 miles, reducing the passage time by approximately 1 hour to just over 3 hours.

The mean tidal stream over this period is:

068°T 1.6 knots

Speed made good 6.0 knots
Time taken 3h 04m
Water track 044°T.
Course 044°T

A.3 **50°51′.2N, 0°42′.7E.**

A.4 **50°53′.6N, 0°47′.2E.**

A.5 (a) **328°C.**

(b) **0836.**

Distance to go 1.9 miles
Speed made good 4 knots

Time taken 0h 29m
Water track 329°T
Course 329°T

A.6 (a) For LW at Rye, no data is available on the height of tide so no predictions can be made. *Reed's Nautical Almanac* gives a drying height on the bar at the entrance to the harbour of 1.5 m above CD. Under the circumstances, a navigator can only be guided by the local sailing directions for the port concerned. For Rye harbour these recommend that without local knowledge **the harbour should only be entered from 1½ hours before HW to HW.**

    (b) **She should watch for coasters leaving or entering the harbour,** and take soundings during entry. The port information shows that tidal balls are hoisted to indicate the depth of water over the bar when coasters are entering or leaving.

    (c) As winds are light, if the boat is unable to enter the harbour she can **anchor inshore** clear of the harbour entrance.

A.7 The VHF radio could have been used to **call Rye Harbour Radio on Channels 16 or 14 to ascertain the depth in the entrance and to check whether the entrance was clear.** (Port radio is operated from 1 hour before to 1 hour after HW.)

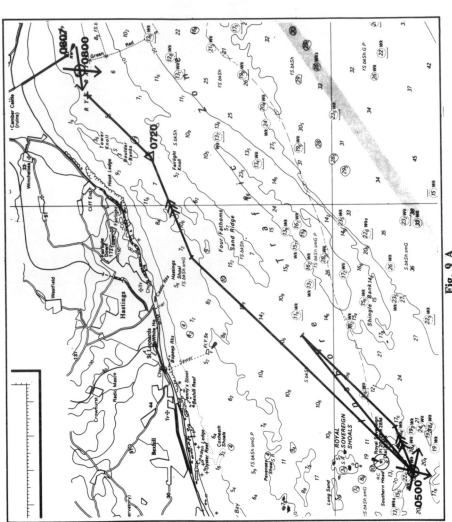

Fig. 9 A

Test Paper B

B.1 (a) **50°23'.6N, 1°28'.4E.**
(b) **50°24'.4N, 1°29'.6E.**
(c) **The disagreement between the EP and the fix suggests that the log is underreading by around 25%.** The boat could not be at the EP. The position is most likely to be on the 1030 position line in the vicinity of the 13 m sounding. The boat should fix her position as soon as possible.
The 0930 fix is 50°20'.3N, 1°29'.5E
Course 352°T
Water track 342°T
Distance run 3 miles
Tidal stream:

042°T 0.6 knots

B.2 (a) **50°29'.8N, 1°29'.9E.**
(b) **50°29'.8N, 1°29'.6E.**
Course 015°T
Water track 005°T
Distance run 4 knots
Tidal stream:

032°T 1.4 knots

B.3 (a) **50°35'.4N, 1°31'.1E**
(b) **The soundings obtained confirm the expected ground track.**
The significant soundings are: 9.8 m, 8.9 m and 12.4 m.
Course 015°T
Water track 005°T
Distance run 4 miles
Speed made good 5.6 knots
Tidal stream:

022°T 1.6 knots

B.4 (a) **001°C.**
(b) **1357.**
Distance to go 8.4 miles
Speed made good 5.8 knots
Time taken 1h 27m
Water track 354°T
Course 004°T
Tidal streams:

007°T 1.7 knots
011°T 2.0 knots (use half)

B.5   **50°44′.4N, 1°30′.2E.**
B.6   **016°T 1.5 miles.**
        Course 030°T
        Water track 030°T
        Distance run 2.7 miles
        Fix 50°48′.2N, 1°33′.0E.
B.7   (a) **188°C**
        (b) **1536**
        Distance to go 3.9 miles
        Speed made good 3.6 knots
        Time taken 1h 05m
        Water track 180°T
        Course 180°T
        Tidal streams:

Tidal diamond N	018°T 3.3 knots
Tidal diamond L	014°T 1.6 knots
This gives a mean of	016°T 2.5 knots

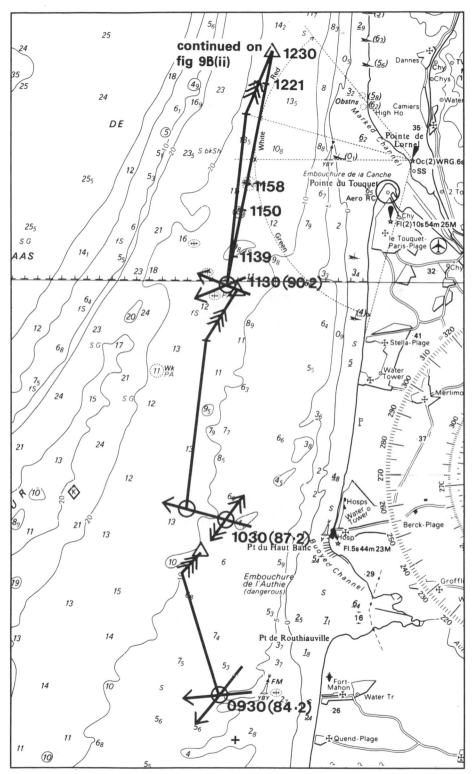

continued on fig 9B(ii)

Fig. 9 B (i)

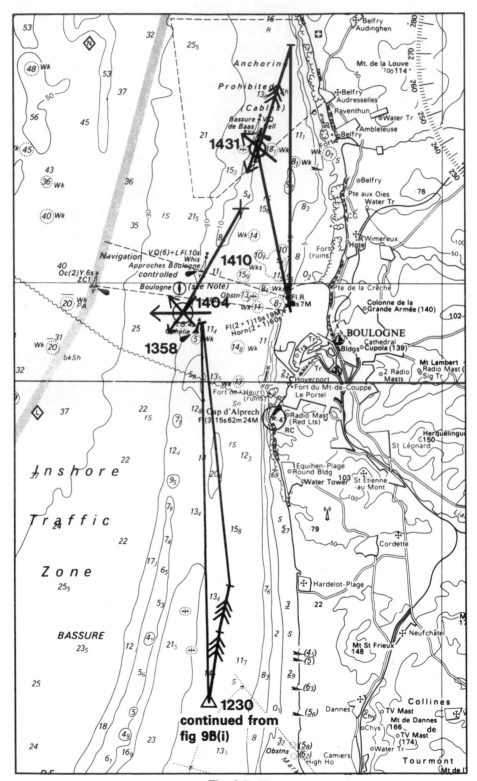

Fig. 9 B (ii)

Test Paper C

C.1 (a) **246°C.**

(b) **1626.**

The distance to go is 30 miles, at 4.5 knots this will take 6 hours 40 minutes (ignoring the tidal streams). Taking into account the tidal streams, which are mostly favourable, the passage time will be reduced to about 6 hours.

The first 2 hours of the tidal stream are foul; the remaining 4 hours fair.

Assuming a 6 hour journey, the mean set and drift will be:

281°T 0.8 knots

A one hour vector diagram using the mean set and drift gives:
Speed made good 5.2 knots
Time taken 5h 46m
Water track 243°T
Course 243°T

C.2 (a) **50°46'.0N, 0°26'.6E.**

(b) **50°46'.0N, 0°26'.6E.**

(c) **Yes:** the EP agrees with the fix.
EP at 1340 50°50'.1N, 0°40'.5E.

C.3 **50°42'.5N, 0°12'.6E.**
Course 245°T
Water track 245°T
Distance run 6.8 miles

C.4 (a) **50°42'.5N, 0°12'.6E.**

(b) **Yes.**

C.5 (a) **312°C.**

(b) **1745.**
Distance to go 7.0 miles
Speed made good 6.5 knots
Time taken 1h 05m
Water track 313°T
Course 313°T

In practice, it would have been sensible to set a course to pass between Royal Sovereign lighthouse and the red buoy to the south of Royal Sovereign Shoals.

Fig. 9 C (i)

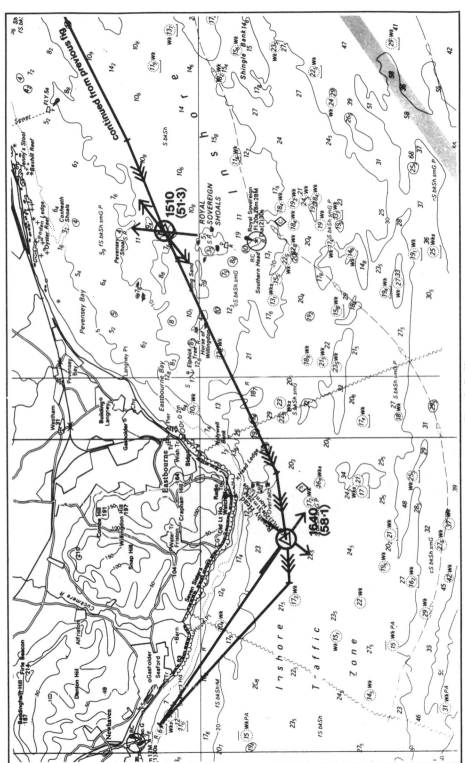

Fig. 9 C (ii)

Test Paper D

D.1   Latest time to leave **0832**.

		HW		LW
	time		height (m)	height (m)
Dover	0304−		5.7+	1.9
Differences	0012		0.4	0.0
Folkestone	0252 GMT		6.1	1.9
	0352 BST			

Range	3.8 m, this is approximately one fifth of the way from neaps to springs.
Height required	1.6 m + 1.0 m + 0.5 m = 3.1 m
Interval	4h 40m after HW

D.2   **0800**.

The distance to go is approximately 21 miles which will take 3h 30m ignoring the tidal streams. The tidal streams become favourable at about 0830. The boat should leave her berth at 0745, aiming to pass the south breakwater at 0800 to allow time for any unforeseen delay that might arise from grounding. The tidal streams from 0800 to 1130 will modify the estimated time to approximately 3h 15m. This would give an ETA of 1115, which will be near LW Rye. There is no data available for Rye on LW times or heights either in *Admiralty Tide Tables* or *Reed's Nautical Almanac*. There is a bar across the entrance which dries 1.5 m. Around LW it is unlikely that there will be sufficient water for the boat to cross the bar. The whole harbour dries well before LW even at neap tides.

High Water Rye

		time	height (m)
High Water Dover	GMT	1524−	5.8+
Differences		0004	0.8
HW Rye	GMT	1520	6.6
	BST	1620	

The ideal time to enter Rye is from 1½ hours to 1 hour before HW (according to the pilotage notes), that is from 1450 to 1520. The boat will have to anchor in Rye Bay and wait for the tide or alternatively she can leave her berth at Folkestone before 0828 and anchor outside the entrance to the fishing harbour until 1130. The forecast wind direction and strength imply that the boat may have to motor from Dungeness.

D.3 (a) **233°C.**
(b) **0952.**
Distance to go 11.6 miles
Speed made good 6.2 knots
Time taken 1h 52m
Water track 223°T
Course 228°T
The tidal streams almost cancel out (0.3 favourable) and as they lie along the track there is no need to plot a vector diagram.

D.4 **Yes.** The position from the buoy is very close to the EP at 0901 and confirms that the boat is making good time.

D.5 The position is **where the position line crosses the 20.4 metre patch.**

D.6 (a) **1059.**
(b) **To starboard.**
(c) **0.2 miles.**
Course 282°T
Water track 282°T
Distance run 6 miles
Speed made good 6.4 knots

D.7 It is a rising tide but the tidal stream is still setting fairly rapidly to the west. The boat should anchor inshore of the red and white buoy out of the tidal stream, and clear of the harbour entrance. Consult a large scale chart and the pilotage notes to ascertain any likely hazards and to find the best anchorage.

General considerations when anchoring:

1. Sufficient chain for at least three times the maximum expected depth; five times if any warp is used.
2. Sufficient water to avoid grounding.
3. Good holding ground.
4. Free of obstructions on the seabed. (If this is not possible the anchor should be buoyed.)
5. Maximum shelter from all expected winds.
6. Clear of all obstructions as the boat swings.
7. Out of areas used frequently by other boats.
8. With suitable shore marks to enable the position to be checked should the anchor drag.
9. Near a suitable landing place.

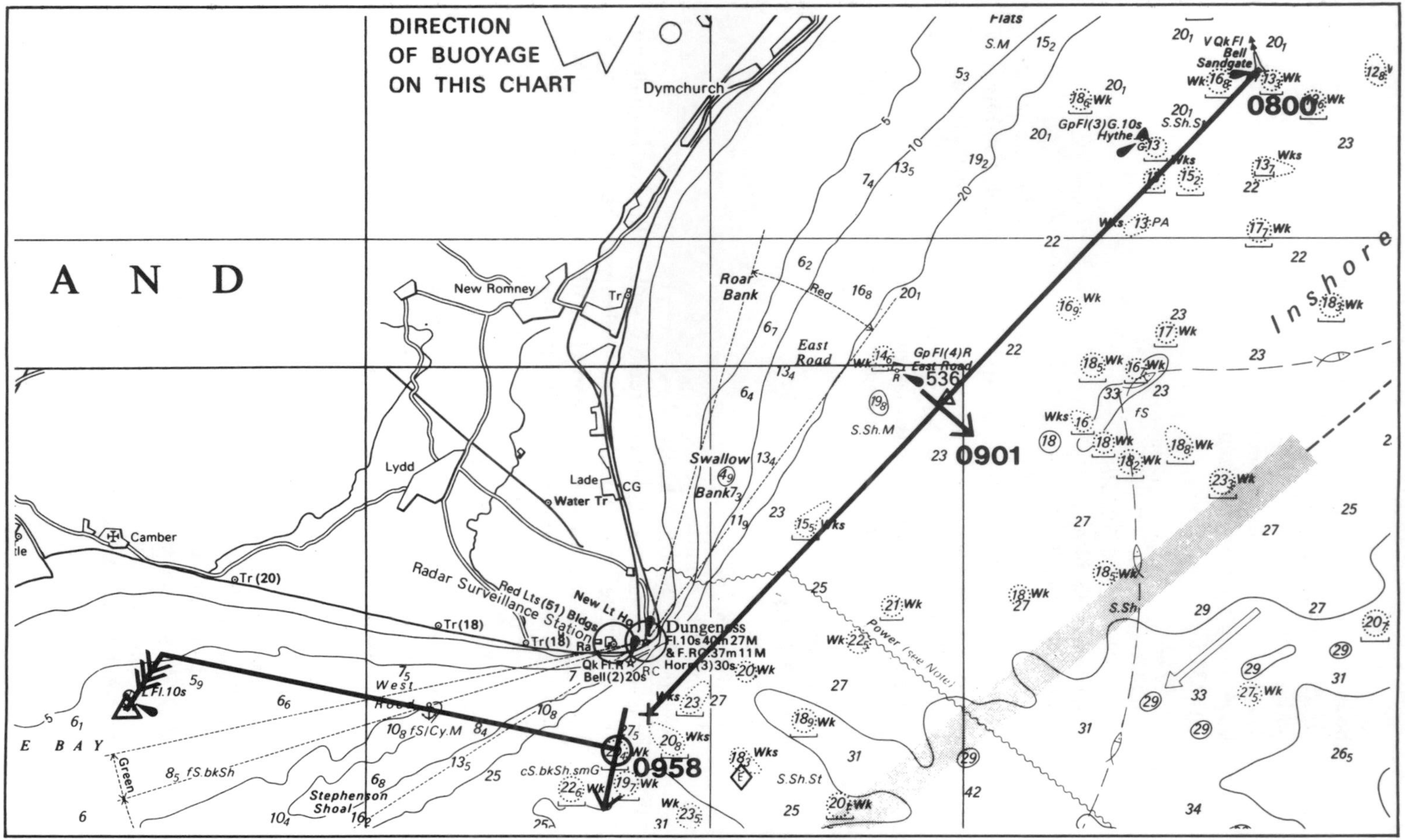
DIRECTION
OF BUOYAGE
ON THIS CHART
A N D
Dymchurch
New Romney
Lydd
Lade
Water Tr
CG
Camber
Tr (20)
Tr (18)
Radar Surveillance Station
Red Lts (51) Bldgs
New Lt Ho
Dungeness
Fl.10s40m27M
& F.R.37m11M
Horn(3)30s
Qk Fl.R
Bell(2)20s
Roar Bank
Red
East Road
Swallow Bank
GpFl(4)R
East Road
GpFl(3)G.10s
Hythe
V Qk Fl
Bell
Sandgate
S.Sh.St
S.Sh.M
Inshore
Power (see Note)
S.Sh
West Road
Stephenson Shoal
L Fl.10s
Green
E BAY
0800
0901
0958
536

Fig. 9 D

Test Paper E

E.1 **Between 1027 and 1127.**

	HW		LW
	time	height (m)	height (m)
Dover	1027 GMT	6.5	0.9
	1127 BST		

Range 5.6 m, a spring tide.
The lock gates open from 1 hour before HW to HW. The tidal stream becomes favourable 4½ hours after HW so she should anchor in the outer harbour until approximately 1530 she can then leave the harbour to catch the first of the west going stream.

E.2 (a) **51°01′.9N, 1°06′.1E.**

(b) **50°57′.7N, 1°00′.8E.**

	Starboard tack	*Port tack*
Course	180°T	270°T
Water track	170°T	280°T

Tidal streams:

1600 – 1700	245°T 0.4 knots
1700 – 1800	224°T 1.6 knots

E.3 **50°53′.9N, 1°00′.8E.**

E.4 (a) **277°C.**

(b) **2041.**

Distance to go 8.0 miles
Speed made good 6.8 knots
Time taken 1h 11m
Water track 287°T
Course 277°T

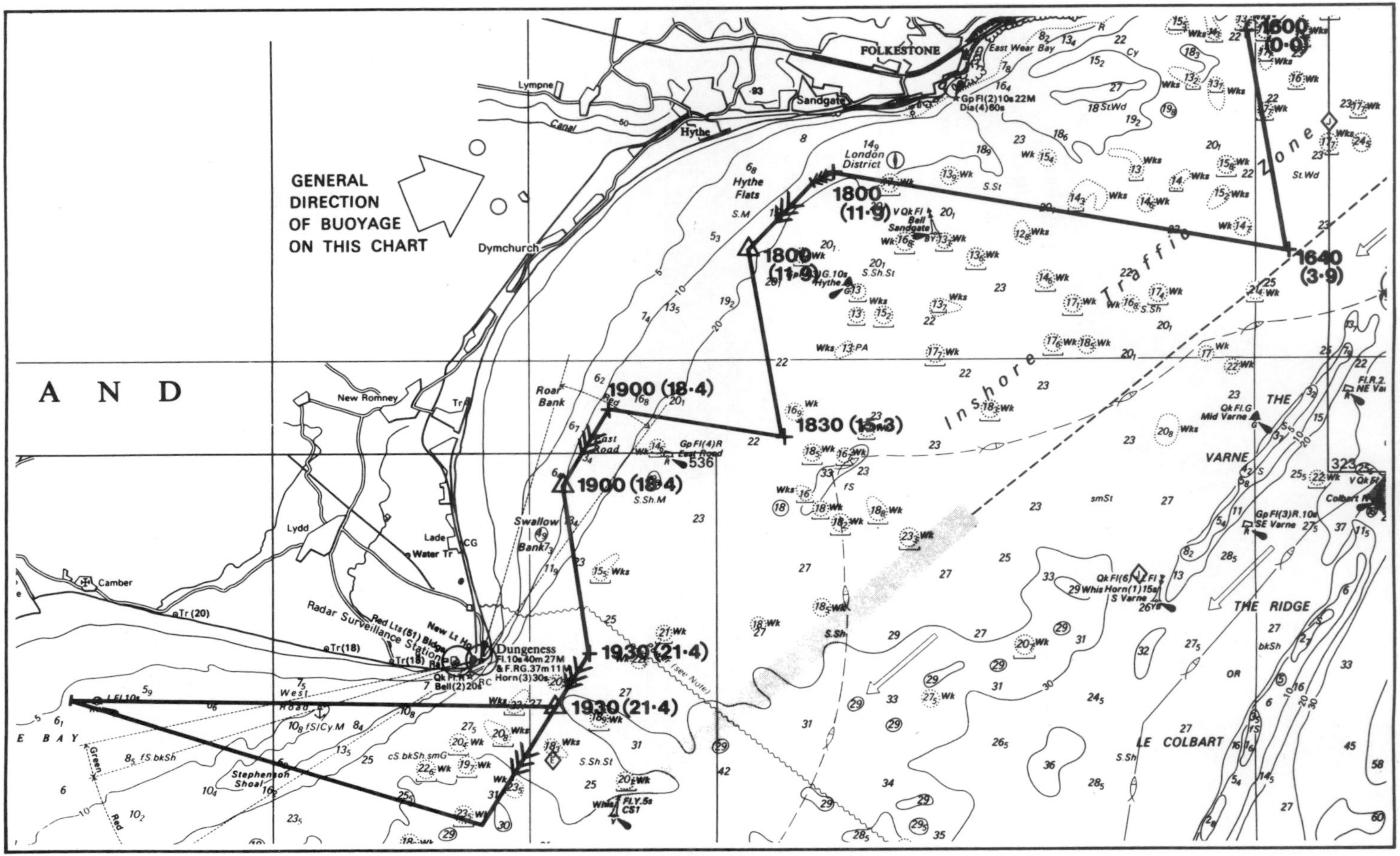
GENERAL
DIRECTION
OF BUOYAGE
ON THIS CHART
A N D
FOLKESTONE
Sandgate
Hythe
Dymchurch
Lympne
Canal
New Romney
Lydd
Camber
Lade
CG
Water Tr
Radar Surveillance Station
Dungeness
Hythe Flats
London District
Roar Bank
Swallow Bank
East Road
East Wear Bay
Stephenson Shoal
West Road
THE VARNE
THE RIDGE
LE COLBART
Inshore Traffic Zone
1600 (0·9)
1640 (3·9)
1800 (11·9)
1800 (11·9)
1830 (15·3)
1900 (18·4)
1900 (18·4)
1930 (21·4)
1930 (21·4)
Fig. 9 E

CHAPTER TEN

Test Paper F

F.1   **50°44'.0N, 1°29'.7E.**

	HW		LW
	time	height (m)	height (m)
Dover	0806 GMT	5.4	2.1
	0906 BST		

Range 3.3 m, it is a neap tide.
Course 348°T
Water track 343°T
Distance run 8.3 miles
Tidal streams:
    011°T 1.1 knots
    014°T 0.9 knots (use half)

F.2   **50°44'.2N, 1°29'.6E.**

F.3   (a) **005°C.**
     (b) **1227.**
     Distance to go 9.4 miles
     Speed made good 6.9 knots
     Time taken 1h 22m
     Water track 001°T
     Course 006°T
     Tidal streams:
        018°T 2.0 knots (use half)
        020°T 1.1 knots (use three quarters)

F.4   **50°55'.6N, 1°27'.2E.**
     Distance run 3.3 miles
     Course 306°T
     Water track 306°T

F.5   **No.** The log is underreading by 0.6 knots. The distance run between 1305 and 1405 was 5.6 miles. As leeway was negligible and the tidal stream slack this gave a speed of 5.6 knots.

F.6   (a) **51°01'.8N, 1°12'.2E.**
     (b) **51°01'.5N, 1°11'.9E.**

F.7   (a) **009°C.**
     (b) **1541.**
     Distance to go 3.0 miles
     Speed made good 5.0 knots
     Time taken 0h 36m
     Water track 005°T
     Course 010°T

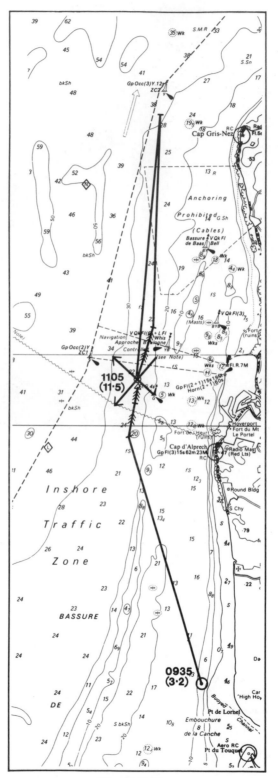

Fig. 10 F (i)

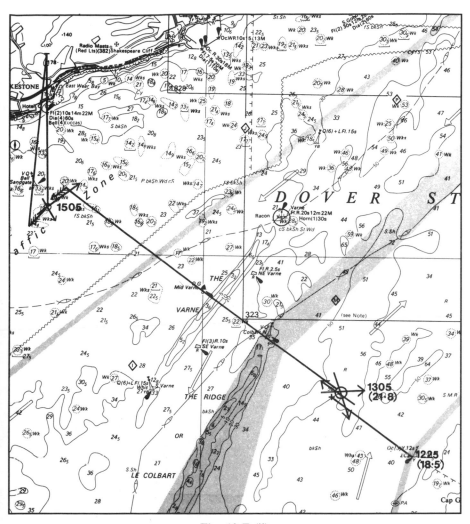

Fig. 10 F (ii)

Test Paper G

G.1   (a) **355°C.**
       (b) **0834.**

	HW			LW
	time	height (m)		height (m)
Dover	2333 GMT	6.7		0.7
	0033 BST			

Range 6.0 m, a spring tide.

	HW	
	time	height (m)
Dover	1200 GMT	6.9
	1300 BST	

Range 6.2 m
Distance to go 38.6 miles which will take 8 h 35 mins making no allowance for the tidal streams.
Tidal streams:
Tidal diamond F

     0033   030°T 1.0 knots
     0133   001°T 0.8 knots
     0233   331°T 0.5 knots
     0333   278°T 0.6 knots

Tidal diamond D

     0433   256°T 0.4 knots
     0533   238°T 0.9 knots
     0633   225°T 1.5 knots

Tidal diamond E

     0700   211°T 1.6 knots
     0800   211°T 2.1 knots

By inspection of the tidal streams the passage will take between 8 and 9 hours.
The mean set and rate of the tidal stream found by plotting 8½ hours of tidal streams are:

     246°T 0.5 knots

Using this to plot an hour vector:
     Speed made good 4.5 knots
     Time taken 8h 35m
     Water track 346°T
     Course 356°T

G.2 **50°28′.0N, 1°03′.8E.**
Course 344°T
Water track 334°T
Distance run 9.3 miles
Tidal streams:

> 030°T 1.0 knot
> 001°T 0.8 knots

G.3 **Yes.** The bearing of 354°T is Dungeness RC. The EP and position line are **within 0.1 miles of each other. The sounding is of little use** for accurate determination of position as the depths across the area arc too similar. The boat's position must be on the position line. The position chosen is that point on the position line which is nearest to the EP.

G.4 **50°33′.7N, 0°59′.4E.**
Course 344°T
Water track 334°T
Distance run 5.4 miles
Tidal stream:

> 331°T 0.5 knots

G.5 **50°44′.8N, 0°52′.2E.**
Course 344°T
Water track 339°T
Distance run 12 miles
Tidal streams:

> Slack
> 256°T 0.4 knots

The sounding is of little use.

G.6 The EP suggests that the boat is still in the traffic separation scheme. **She should carry on on the same course and fix her position as soon as possible**.

G.7 (a) **50°47′.4N, 0°50′.0E.**
(b) **50°47′.9N, 0°49′.8E.**
Course 344°T
Water track 344°T
Distance run 3 miles
Tidal stream:

> 238°T 0.9 knots (use half)

G.8 (a) **No.** The onshore wind will make conditions uncomfortable and dangerous. If the wind increases further the anchor will probably drag causing the boat to ground on a lee shore.

(b) With the weather report and the expected arrival at Rye at
LW, it would have been better to make an initial decision to
aim for a port where access does not depend upon tidal
conditions before entering the traffic separation scheme. The
choices available are **Folkestone**, which also partially dries, or
Newhaven which is accessible at any state of tide. Distance to
go is about the same to either port. Inspection of the tidal
streams suggests that a foul tidal stream will be experienced
all the way to Folkestone. There will be a fair tidal stream for
the first two hours towards Newhaven. Newhaven is chosen.

G.9 **50°40′.7N, 0°21′.2E.**
Course 244°T
Water track 249°T
Distance run 18 miles

G.10 **50°41′.5N, 0°18′.2E.**
Course 273°T
Water track 278°T
Distance run 2.8 miles

G.11 (a) **282°C.**
(b) **1125.**
Distance to go 10.6 miles
Speed made good 4.4 knots
Time taken 2h 25m
Water track 287°T
Course 282°T

G.12 **The continuous sounding of a bell from the west pier.** On the
approach to Newhaven the west pier should be identified so as to
locate the signal mast. A red ball indicates entry and departure
permitted by small vessels.

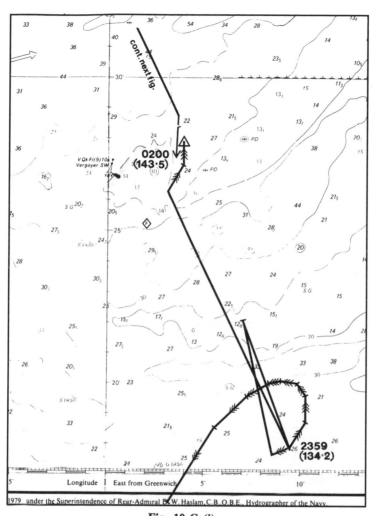

Fig. 10 G (i)

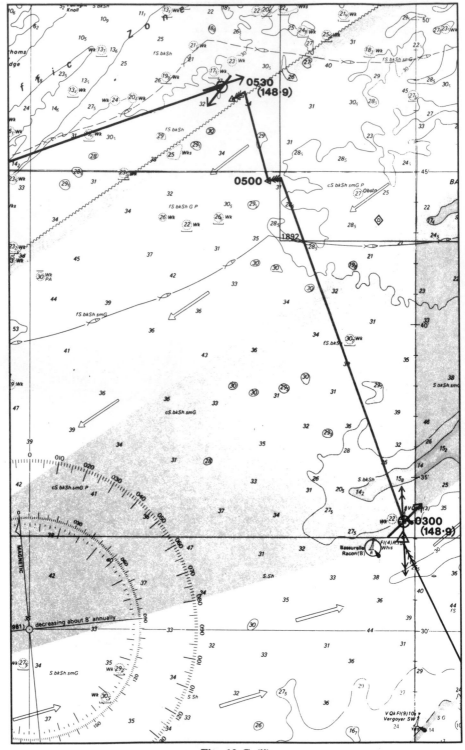

Fig. 10 G (ii)

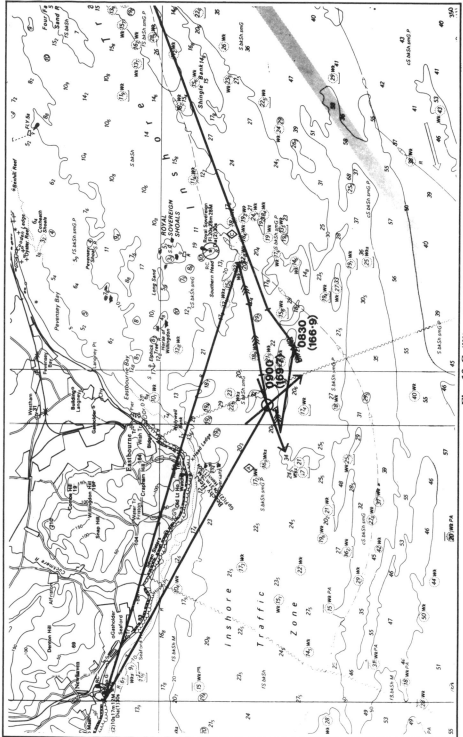

Fig. 10 G (iii)

Test Paper H

H.1   **50°22′.3N,1°27′.1E.**
       The light is Cap d'Alprech.

Height of light above MHWS (m)	62.0+
Height of MHWS Boulogne (m)	8.9
	───
Height of light above CD (m)	70.9−
Height of tide (m)	1.1
	───
Height of light above sea level (m)	69.8
	───

Distance off 20 miles.

H.2   (a) **50°28′.3N, 1°22′.8E.**
       ((b) **50°28′.3N, 1°22′.6E.**
       (c) **The fix is within 0.2 miles of the EP.**
       Course 000°T
       Water track 350°T
       Distance run 9.2 miles
       Tidal streams:

           203°T 1.7 knots
           197°T 1.6 knots

       HW Dover on the 13 August is at 0255 and it is a spring tide.
H.3   (a) **2348.**
       (b) **0325.**
       A 1 hour vector diagram has been plotted on port tack and a 3½
       hour vector diagram on starboard tack to determine the ground
       tracks.
       Port tack
           Course 095°T
           Water track 105°T
           Ground track 114°T
           Speed made good 4.6 knots
           Tidal stream:

               187°T 0.7 knots

       Starboard tack
           Course 355°T
           Water track 345°T
           Ground track 352°T
           Speed made good 5.3 knots
           Tidal streams:
           Tidal diamond K

042°T 0.6 knots
032°T 1.4 knots (use half)

Tidal diamond L

076°T 0.5 knots (use half)
019°T 1.0 knot
007°T 1.7 knots (use half)

H.4 **See plot.**
H.5 **50°35′.4N, 1°29′.8E.**
Course 355°T
Water track 345°T
Distance run from 2359 2.5 miles
Tidal stream from 2359:

042°T 0.6 knots (use half)

Distance run from 0030 4.4 miles
Tidal stream from 0030:

032°T 1.4 knots

H.6 (a) **50°40′.3N, 1°28′.6E.**
(b) Set and drift **022°T 1.0 mile**
Distance run 4.1 miles
H.7 **50°47′.5N, 1° 21′.2E.**
Course from 0255 303°T
Water track 298°T
Distance run 7.2 miles
Tidal streams:

007°T 1.7 knots
011°T 2.0 knots (use half)

H.8 **Yes;** the soundings agree with the 0500 EP
Speed made good 5.2 knots
Tidal streams:

025°T 1.4 knots (use half)
034°T 1.2 knots (use half)

EP 50°51′.0N, 1°15′.2E
H.9 **0619.**
Distance to go 6.2 miles
Speed made good 4.7 knots
II.10 **50°55′.7N, 1°08′.6E.**
The fix indicates that the boat has cleared the traffic separation scheme.

H.11 (a) **006°C.**
      (b) **0747.**
      Distance to go 9.9 miles
      Speed made good 6.8 knots
      Times taken 1h 22m
      Water track 008°T
      Course 008°T

H.12 (a) **No.** She cannot enter her berth.
      (b) She must anchor and wait until **1214**

	HW		LW	HW	
	time	height (m)	height (m)	time	height (m)
Dover	0155− GMT	6.6+	0.9−	1413− GMT	6.7+
Differences	0015	0.4	0.1	0014	0.4
Folkestone	0140	7.0	0.8	1359	7.1
	0240 BST			1459 BST	

Ranges   5.7 m and 5.8 m, spring tides
Height required   3.0 m
Interval   4h 25m after HW and 2h 45m before HW
The boat cannot reach her berth between 0705 and 1214.

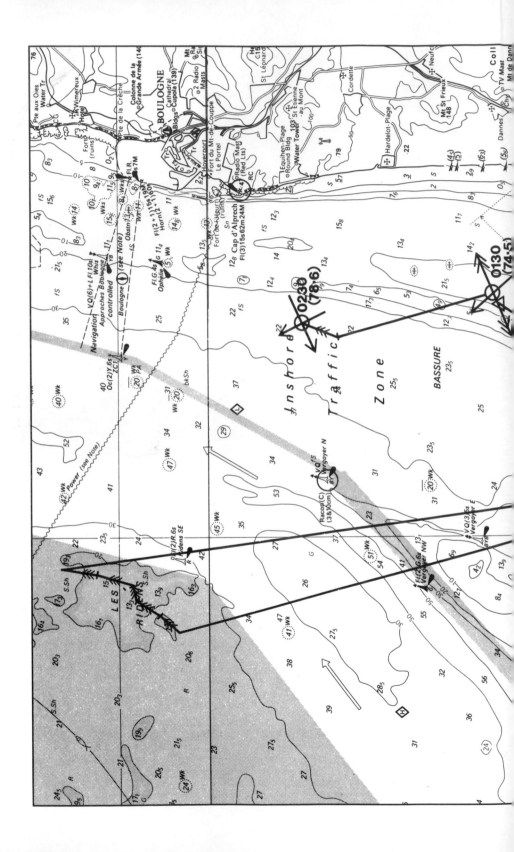

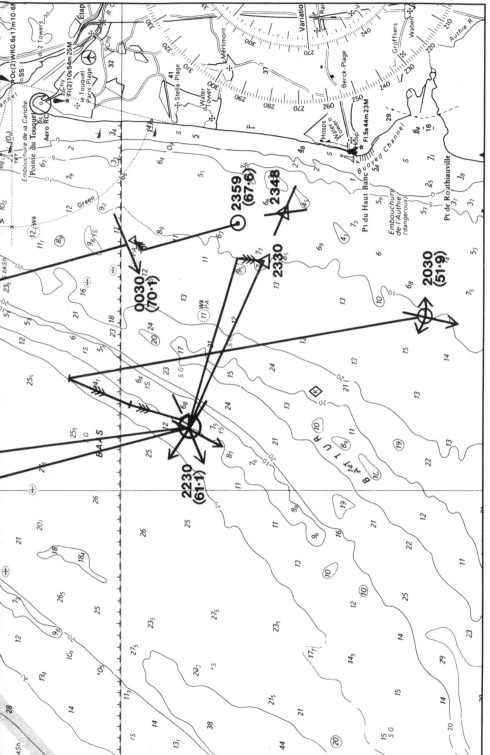

Fig. 10 H (i)

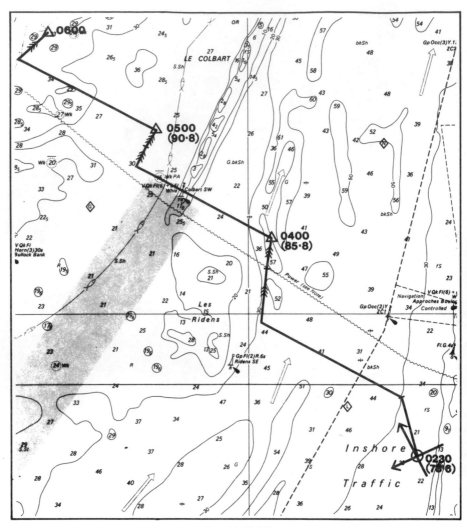

Fig. 10 H (ii)

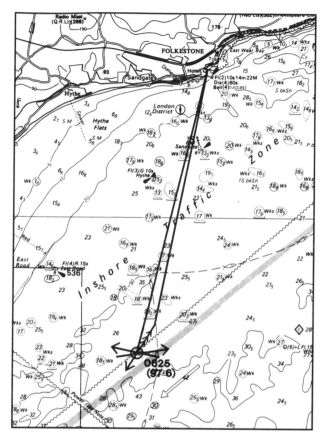

Fig. 10 H (iii)

Test Paper 1

I.1   The time to leave has to be between 1 hour before HW and HW:
       0604 BST to 0704 BST.

       1. Obtain a weather report at 0555.
       2. Check all gear, supplies and fuel.
       3. See that everything is well stowed in case of foul weather.
       4. Check safety equipment. See that every member of the crew
          knows where it is stowed and how to use it. Issue safety
          harnesses and life jackets to each crew member so that
          they can adjust these ready for use if necessary.
       5. Plot the first leg of the passage and decide whether the
          passage should be undertaken in one go.
       6. Measure the total distance to go and estimate the passage
          time. Work out whether the boat can enter her berth
          upon arrival or whether she will need to go to a
          temporary alternative berth. Check for any local
          anchorages that will be sheltered in the conditions
          forecast.
       7. Check if there are any intermediate ports of refuge in case of
          bad weather.
       8. Make a list of landmarks that can be identified on the chart
          which are suitable for bearings. List radio direction
          finding stations giving their identification signal, broad-
          cast time, operating frequency, and range.
       9. Decide upon a suitable point to cross the traffic separation
          scheme at right angles.
      10. Note the VHF channels used for port entry. Brief the crew on
          Mayday procedure.
      11. Leave word with a responsible person ashore concerning the
          proposed passage and ETA.
      12. Always take into account the seaworthiness of the boat and
          the ability of the crew and do not hesitate to cancel or
          shorten the passage if in doubt about weather con-
          ditions.
      13. Keep regular entries in the log book.

I.2   (a) **353°C.**
       (b) **0907.**
       Distance to go 8.7 miles
       Speed made good 6.0 knots
       Time taken 1h 27m
       Water track 359°T
       Course 354°T
       Tidal streams:

011°T 1.1 knots
014°T 0.9 knots (use half)

HW Dover on 4 August is 0708, it is a neap tide.

I.3   See Extract 16. Normal day signal for entry:
   (a) **2 cones points together over a cone point down (vertically disposed); or sometimes just a green flag.**
   (b) **From the west entrance to Darse Sarreg-Bournet.**

I.4   **0722.**

	time	HW height (m)	LW height (m)
Dover	0721– GMT	5.5+	2.0
Differences	0009	0.4	0.0
Folkestone	0712	5.9	2.0

0812 BST

Range	3.5 m, it is a neap tide
Height required	3.1 m
Interval	4h 40m after HW

The boat must reach her berth by 1252. The passage time is approximately 5 hours. Allowing 30 minutes to clear the harbour, the latest time to start from Boulogne will be 0722. To allow for delays it is best to start 1 hour before this time. The boat should aim to be at the entrance no later than 0630 leaving her berth at 0600.

I.5   **See plot.**
I.6   (a) **297°C.**
       (b) **0813.**
Distance to go 4.8 miles
Speed made good 5.4 knots
Time taken 0h 53m
Water track 302°T
Course 297°T
Tidal streams:

   014°T 0.6 knots (use half)
   017°T 1.6 knots (use half)

I.7   **50°49′.2N, 1°28′.6E.**
I.8   **50°54′.0N, 1°23′.4E.**
Course 303°T
Water track 308°T
Distance run 4.9 miles
Tidal streams:

017°T 1.6 knots (use half)
018°T 2.2 knots (use half)

I.9   (a) **50°56′.2N, 1°20′.7E.**
     (b) **018°T 2.6 miles.**

I.10   **1040.**
Distance to go 4.2 miles
Speed over the ground 4.9 knots
Tidal stream:

    052°T 1.0 knot

I.11   (a) **51°00′.0N, 1°15′.8E.**
     (b) **Two position lines will not show any error**, but the fix is near the EP and there is no immediate danger as the boat is leaving the traffic separation scheme.

I.12   (a) **313°C.**
     (b) **1154.**
Distance to go 5.2 miles
Speed made good 4.9 knots
Time taken 1h 04m
Water track 319°T
Course 314°T
Tidal stream:

    061°T 1.0 knot

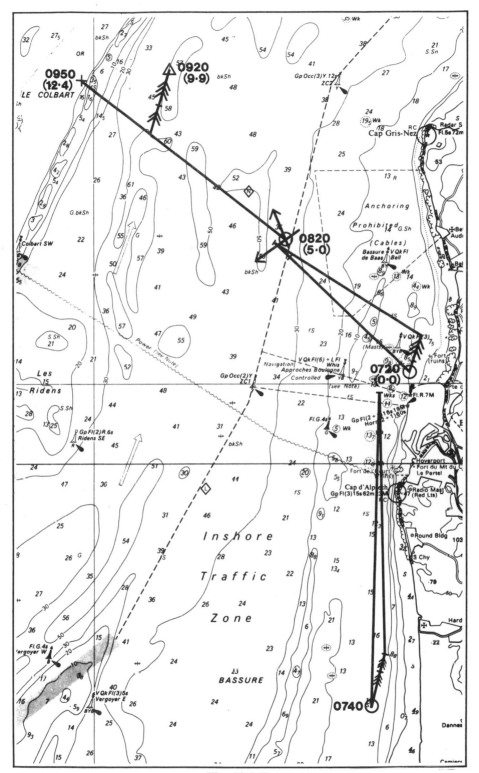

Fig. 10 I (i)

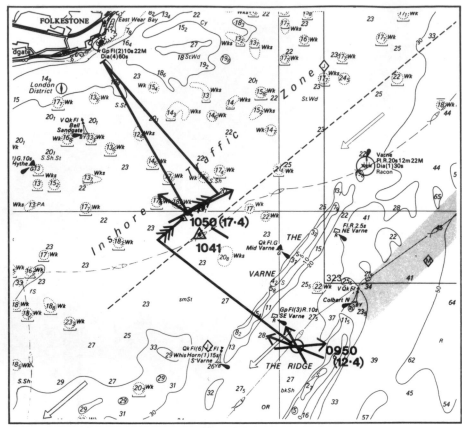

Fig. 10 I (ii)

Test Paper J

J.1   **1345.** HW Dover is at 0255 and 1513, spring tides. HW Rye is at
      1512. Off Calais the tidal stream is slack, becoming west going at
      0700. The preferred time to arrive at Rye entrance is from 1½
      hours to 1 hour before HW that is, from 1342 to 1412. From the
      RW buoy to the entrance will take about 20 minutes which gives a
      latest ETA at the buoy of around 1345. With approximately 45
      miles to go from Calais of which about 9 miles can be accounted
      for by favourable tidal streams, the passage time at 7 knots should
      be in the region of 5 hours, giving a latest starting time of 0845, so
      0700 would be a sensible time to start.

J.2   (a) **240°C.**
      (b) **0942.**
      Distance to go 4.8 miles
      Speed made good 9.2 knots
      Time taken 0h 31m
      Water track 236°T
      Course 236°T
      Tidal stream:

          243°T 2.3 knots (use half)

J.3   **1139.**
      The 1044 EP has been plotted
      Course 316°T
      Water track 316°T
      Distance to go from the 1044 EP 6.0 miles
      Speed made good from 1044 to 1144 6.6 knots
      Tidal streams:

          204°T 2.5 knots
          208°T 2.7 knots

J.4   **Fix the position** and make every attempt to sail if any wind
      appears. It would be prudent to **call Dover Radar Surveillance** on
      VHF Channel 10 giving the boat's estimated position. If repairs
      cannot be effected, assistance may be requested through Dover
      Coastguard (Channel 16). Inspect the tidal stream and **estimate the
      drift. Keep a good look out.**

J.5   **51°00′.4N, 1°20′.4E.**

J.6   The boat should **motor clear of the traffic separation scheme**, fixing
      her position as she does so. Dover Radar Surveillance should be
      informed when the boat is clear.

J.7   (a) **51°02′.0N, 1°14′.3E.**
      (b) **51°01′.8N, 1°14′.7E.**

J.8   (a) **233°C.**
       (b) **1438.**
       Distance to go 12.9 miles
       Speed made good 6.8 knots
       Time taken 1h 54m
       Water track 229°T
       Course 229°T

J.9   (a) **50°55′.4N, 1°01′.8E.**
       (b) The boat is well behind schedule and will not reach Rye in
           good time. She could **proceed due north and anchor** in a
           suitable depth of water in the vicinity of East Road where she
           could wait for visibility to improve. It would not be
           unreasonable to **use the echo sounder to follow the 10 m depth
           contour in a north easterly direction towards Folkestone or
           even Dover**. This would also be with a favourable tidal stream.

Fig. 10 J (i)

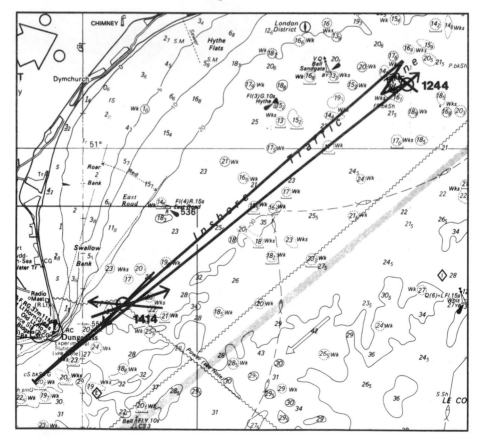

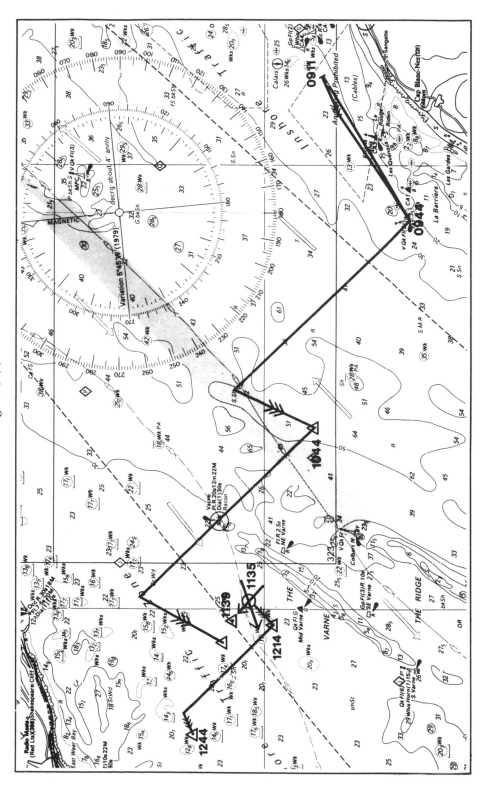

Fig. 10 J (ii)

Extracts and Data for Exercises

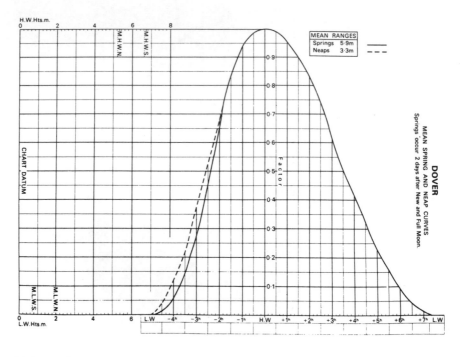

Extract 1a

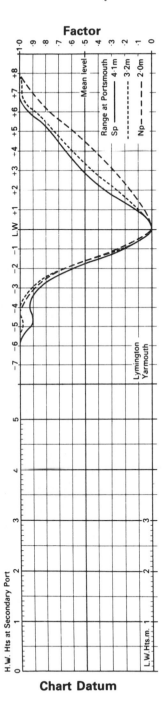

Chart Datum

Extract 1b

Extract 2   *Extract 3*

DOVER
HIGH & LOW WATER

AUGUST

GMT

Day	Time h.min.	Ht. m.		Time h.min.	Ht. m.
1 M	0308	5.7	**16** Tu	0423	5.7
	1013	1.8		1136	1.9
	1524	5.9		1649	5.8
	2244	1.7		—	—
2 Tu	0402	5.5	**17** W	0015	1.9
	1059	2.0		0533	5.4
	1619	5.7		1238	2.2
	2336	1.9		1805	5.5
3 W	0502	5.4	**18** Th	0127	2.1
	1157	2.1		0655	5.3
	1722	5.6		1401	2.2
	—	—		1926	5.4
4 Th	0041	1.9	**19** F	0246	2.1
	0608	5.4		0806	5.4
	1310	2.2		1518	2.1
	1832	5.5		2034	5.5
5 F	0154	1.9	**20** Sa	0350	1.9
	0721	5.5		0905	5.7
	1430	2.0		1616	1.8
	1945	5.7		2131	5.7
6 Sa	0314	1.7	**21** Su	0441	1.7
	0834	5.8		0952	6.0
	1552	1.6		1704	1.6
	2053	5.9		2214	5.9
7 Su	0430	1.4	**22** M	0525	1.5
	0935	6.1		1030	6.2
	1659	1.3		1746	1.4
	2152	6.2		2249	6.1
8 M	0533	1.1	**23** Tu	0604	1.4
	1027	6.5		1102	6.4
	1758	0.9		1824	1.2
	2244	6.5		2318	6.2
9 Tu	0634	0.9	**24** W	0639	1.2
	1115	6.7		1133	6.5
	1855	0.7		1859	1.1
	2333	6.7		2346	6.3
10 W	0728	0.7	**25** Th	0712	1.2
	1200	6.9		1203	6.5
	1947	0.5		1933	1.1
	—	—		—	—
11 Th	0022	6.8	**26** F	0014	6.3
	0818	0.6		0742	1.2
	1245	6.9		1232	6.5
	2034	0.4		2004	1.1
12 F	0109	6.7	**27** Sa	0045	6.3
	0901	0.7		0809	1.2
	1330	6.9		1302	6.5
	2118	0.5		2032	1.1
13 Sa	0155	6.6	**28** Su	0116	6.2
	0939	0.9		0836	1.3
	1413	6.7		1330	6.4
	2157	0.7		2058	1.2
14 Su	0240	6.3	**29** M	0145	6.1
	1014	1.1		0905	1.4
	1458	6.5		1358	6.2
	2238	1.1		2131	1.4
15 M	0328	6.1	**30** Tu	0219	5.9
	1051	1.5		0941	1.6
	1549	6.1		1433	6.0
	2322	1.5		2207	1.6
			31 W	0304	5.7
				1023	1.9
				1524	5.8
				2257	1.9

PORTSMOUTH
HIGH & LOW WATER

APRIL

GMT

Day	Time h.min.	Ht. m.		Time h.min.	Ht. m.
1 F	0131	4.8	**16** Sa	0101	4.7
	0639	0.6		0615	0.5
	1349	4.6		1322	4.5
	1858	0.8		1834	0.6
2 Sa	0205	4.7	**17** Su	0141	4.6
	0714	0.8		0652	0.5
	1428	4.5		1405	4.4
	1931	1.1		1912	0.8
3 Su	0238	4.5	**18** M	0222	4.4
	0751	1.1		0733	0.7
	1507	4.2		1452	4.3
	2010	1.4		1955	1.1
4 M	0316	4.2	**19** Tu	0308	4.2
	0835	1.4		0821	1.0
	1554	3.9		1545	4.0
	2101	1.8		2050	1.4
5 Tu	0405	3.9	**20** W	0405	3.9
	0932	1.7		0924	1.5
	1655	3.7		1655	3.8
	2212	2.0		2208	1.7
6 W	0513	3.6	**21** Th	0522	3.7
	1051	2.0		1049	1.5
	1822	3.5		1826	3.8
	2346	2.1		2342	1.7
7 Th	0644	3.4	**22** F	0652	3.7
	1224	1.9		1220	1.4
	1950	3.5		1949	4.0
	—	—		—	—
8 F	0111	1.9	**23** Sa	0104	1.5
	0807	3.5		0809	3.9
	1337	1.7		1332	1.1
	2054	3.8		2052	4.2
9 Sa	0211	1.6	**24** Su	0206	1.2
	0905	3.8		0907	4.2
	1427	1.4		1428	0.8
	2138	4.0		2141	4.5
10 Su	0252	1.3	**25** M	0256	0.9
	0948	4.0		0955	4.4
	1504	1.1		1516	0.6
	2214	4.3		2225	4.7

Extract 4

TIME AND HEIGHT DIFFERENCES – SECONDARY PORTS

			N.	E.			0000 and 1200	0600 and 1800	0100 and 1300	0700 and 1900	6·7	5·3	2·0	0·8
83	Newhaven	. . .	50 47	0 04	−0010	−0015	0000	0000	+0·4	+0·2	0·0	−0·2		
84	Eastbourne	. . .	50 46	0 17	−0005	−0010	+0015	+0020	+1·1	+0·6	+0·2	+0·1		
89	**DOVER**	. . . .	(see page 22)						6·7	5·3	2·0	0·8		
85	Hastings	. .	50 51	0 35	0000	0010	−0030	−0030	+0·8	+0·5	+0·1	−0·1		
86	Rye (Approaches)	.	50 55	0 47	+0005	−0010	⊚	⊚	+1·0	+0·7	⊚	⊚		
86a	Rye (Harbour)	.	50 56	0 46	+0005	−0010	⊚	⊚	−1·4	−1·7	⊚	⊚		
87	Dungeness	. .	50 54	0 58	−0010	−0015	−0020	−0010	+1·0	+0·5	+0·2	−0·1		
88	Folkestone	. .	51 05	1 12	−0020	−0005	−0010	−0010	+0·4	+0·4	0·0	−0·1		
89	**DOVER**	. . . .	51 07	1 19	STANDARD PORT				See Table V					
98	Deal	. . .	51 13	1 25	+0010	+0020	+0010	+0005	−0·6	−0·3	0·0	0·0		
102	Ramsgate	. .	51 20	1 25	+0020	+0020	−0007	−0007	−1·8	−1·5	−0·8	−0·4		

Extract 5

TIDAL DIFFERENCES ON DOVER

PLACE	MHW		MLW		GUIDING DEPTH AT			
	Tm. Diff.	Ht. Diff.	Tm. Diff.	Ht. Diff.	HWS	HWN	CD	POSITION
	h. min.	m.	h. min.	m.	m.	m.	m.	
Hastings	− 0 05	+0.6	− 0 30	0.0	9.0	7.3	1.5	Entrance
Rye (Apprs.)	0 00	+0.8	−	−	6.2	4.5	− 1.5	Bar near entrance
Dungeness	− 0 15	+1.2	− 0 15	+0.2	15.3	13.6	7.3	West Road Anche.
Folkestone	− 0 10	+0.4	− 0 10	0.0	5.5	3.1	− 1.6	Alongside Sth Quay
Dover	0 00	0.0	0 00	0.0	7.1	5.7	0.4	Entce. Granville Dock
Deal	+ 0 15	− 0.4	+ 0 05	0.0	10.1	9.0	4.0	Pier Head
Richborough	+ 0 15	− 1.0	−	−	2.8	1.7	− 0.9	Chan to
Ramsgate	+ 0 20	− 1.6	− 0 07	− 0.6	5.0	3.9	0.1	Entrance

Extract 6

TIDAL DIFFERENCES ON PORTSMOUTH

Newport	−	− 0.5	−	+ 0.7	2.3	1.6	− 1.8	
Newtown Creek	− 0 50S / + 0 10N	− 1.1	− 0 15	− 0.1	4.3	3.7	0.7	Off Entce. to Creek
Yarmouth	− 1 05S / + 0 05N	− 1.4	− 0 25	− 0.2	5.0	4.4	1.9	Castle Pier
Totland Bay	− 1 30S / − 0 45N	− 1.8	− 0 40	− 0.3	4.3	3.9	1.6	Pier head
Alum Bay	− 1 40S / − 0 50N	− 1.8	− 0 40	− 0.3	−	−	−	Anchorage prohibited
The Solent Hurst Point	− 1 16S / − 0 05N	− 1.7	− 0 25	− 0.3	10.7	10.3	8.0	Hurst Road
Keyhaven	− 1 05S / 0 00N	− 1.6	− 0 25	− 0.3	6.6	6.1	3.7	In Lake behind bar
Lymington	− 0 55S / + 0 05N	− 1.5	− 0 20	− 0.3	4.4	4.0	1.4	Bar
Beaulieu River ...	− 0 25S / + 0 05N	− 0.6	− 0 15	− 0.1	4.6	3.8	0.6	Bar
Calshot Castle ..	+ 0 10	− 0.2	− 0 15	0.0	17.0	16.3	12.6	Channel
Lee-on-Solent	0 00	− 0.1	− 0 10	+ 0.1	8.4	7.7	4.0	North Channel

Extract 7

TABLE FOR FINDING DISTANCE OFF WITH SEXTANT UP TO 6 MILES

Distance in Miles & Cables	HEIGHT OF OBJECT, TOP LINE METRES—LOWER LINE FEET												Distance in Miles & Cables
	12 40	15 50	18 60	21 70	24 80	27 90	30 100	33 110	37 120	40 130	43 140	46 150	
m c	° ′	° ′	° ′	° ′	° ′	° ′	° ′	° ′	° ′	° ′	° ′	° ′	**m c**
0 6	0 38	0 47	0 57	1 06	1 15	1 25	1 34	1 44	1 53	2 02	2 12	2 21	**0 6**
0 7	0 32	0 40	0 48	0 57	1 05	1 13	1 21	1 29	1 37	1 45	1 53	2 01	**0 7**
0 8	0 28	0 35	0 42	0 49	0 57	1 04	1 11	1 18	1 25	1 32	1 39	1 46	**0 8**
0 9	0 25	0 31	0 38	0 44	0 50	0 57	1 03	1 09	1 15	1 22	1 28	1 34	**0 9**
1 0	0 23	0 28	0 34	0 40	0 45	0 51	0 57	1 02	1 08	1 14	1 19	1 25	**1 0**
1 1	0 21	0 26	0 31	0 36	0 41	0 46	0 51	0 57	1 02	1 07	1 12	1 17	**1 1**
1 2	0 19	0 24	0 28	0 33	0 38	0 42	0 47	0 52	0 57	1 01	1 06	1 11	**1 2**
1 3	0 17	0 22	0 26	0 30	0 35	0 39	0 44	0 48	0 52	0 57	1 01	1 05	**1 3**
1 4	0 16	0 20	0 24	0 28	0 32	0 36	0 40	0 44	0 48	0 53	0 57	1 01	**1 4**
1 5	0 15	0 19	0 23	0 26	0 30	0 34	0 38	0 41	0 45	0 49	0 53	0 57	**1 5**

Extract 8

TABLE II—TO FIND DISTANCE OFF LIGHTS RISING OR DIPPING

Height of Light		HEIGHT OF EYE												
		Metres												
		1.5	3	4.6	6.1	7.6	9.1	10.7	12.2	13.7	15.2	16.8	18.3	19.8
		Feet												
		5	10	15	20	25	30	35	40	45	50	55	60	65
m	ft													
58	**190**	18½	19½	20¼	21	21½	22	22¾	23	23½	24	24½	24¾	25
61	**200**	18¾	20	20¾	21½	22	22½	23	23½	24	24½	24¾	25¼	25½
64	**210**	19¼	20¼	21	21¾	22½	23	23½	24	24½	24¾	25¼	25½	26
67	**220**	19½	20¾	21½	22¼	22¾	23¼	24	24¼	24¾	25¼	25½	26	26¼
70	**230**	20	21	22	22½	23¼	23¾	24¼	24¾	25	25½	26	26¼	26¾
73	**240**	20½	21½	22¼	23	23½	24	24½	25	25½	26	26¼	26¾	27
76	**250**	20¾	21¾	22½	23¼	24	24½	25	25½	26	26¼	26¾	27	27½
79	**260**	21	22¼	23	23¾	24¼	24¾	25¼	25¾	26¼	26¾	27	27½	27¾
82	**270**	21½	22½	23¼	24	24½	25¼	25¾	26¼	26½	27	27½	27¾	28¼
85	**280**	21¾	23	23¾	24½	25	25½	26	26½	27	27½	27¾	28	28½
88	**290**	22	23¼	24	24¾	25¼	26	26½	26¾	27¼	27¾	28	28½	28¾
91	**300**	22½	23½	24½	25	25¾	26¼	26¾	27¼	27½	28	28½	28¾	29¼
95	**310**	22¾	24	24¾	25½	26	26½	27	27½	28	28½	28¾	29	29½
98	**320**	23	24¼	25	25¾	26¼	27	27½	27¾	28¼	28¾	29	29½	29¾
100	**330**	23½	24½	25¼	26	26½	27¼	27¾	28	28½	29	29½	29¾	30
104	**340**	23¾	24¾	25¾	26¼	27	27½	28	28½	29	29¼	29¾	30	30½
107	**350**	24	25	26	26¾	27¼	27¾	28¼	28¾	29¼	29½	30	30½	30¾
122	**400**	25½	26½	27½	28	28¾	29¼	29¾	30¼	30¾	31	31½	32	32¼
137	**450**	27	28	28¾	29½	30	30¾	31¼	31¾	32	32½	33	33¼	33¾

Extract 9

MARINE AND AERONAUTICAL RADIOBEACONS

REED'S STN. No.	STATION NAME AND GROUPING	LAT. North ° '	LONG West ° '	RANGE AND FREQ.		MORSE IDENT.	TRANSMISSION Times/sequence Modes and FOOTNOTES	
90	Newhaven	50 47	0 03½ E	10	303.4	NH –· ····	3, 6	A2A
93	Poole Harbour	50 41	1 57	10	303.4	PO ·––· –––	3, 0	A2A
105	Royal Sovereign Lt.	50 43½	0 26	50	310.3	RY ·–· –·––	2	A2A
108	Pointe d'Ailly	49 55	0 57½	50	310.3	AL ·– ·–··	3	A2A
111	Cap d'Alprech Lt.	50 42	1 34	20	310.3	PH ·––· ····	5	A2A
114	Cap Griz Nez Lt. Ho.	50 52	1 35	30	310.3	GN ––· –·	1, 4	A2A
117	Dungeness Lt.	50 55	0 58½	30	310.3	DU –·· ··–	6	A2A
A117	Le Touquet	50 32	1 35½	20	358	LT ·–·· –	Cont.	A2A
135	South Foreland Lt.	51 08½	1 22½	30	305.7	SD ··· –··	6	A2A

Extract 10

NEWHAVEN 50°46'N 0°04'E. Tel: Newhaven 4131 (4132).
Pilotage: T.H. Compulsory with certain exceptions. E.T.A. 12 h in advance via GPO Coast Station and 2 h in advance via Newhaven Harbour Radio.
P/Station: W. Pier Head.
Radio—Port: VHF Chan. 16, 12—continuous.
Entry Signals: West Pier Lt. H.—for bells sounded cont. }   cross chan. steamers
                East Pier Inner.—G. Lt.   }   expected or leaving.
                Newhaven. (Mast at base of Lt. Ho. West Pier).

Extract 11

NEWHAVEN   Chart BA 2154
HW +0.00. DS in offing: E −5.35; W −0.15
MHWS 6.6; MLWS 0.5; MTL 3.6; MHWN 5.2; MLWN 1.9.
The hr is accessible in all weathers. There is 4.2 m in the entrance which is protected by a breakwater 715m long. The hr lies immediately E of Burrow Hd which is the E end of the line of cliffs stretching from Brighton to Newhaven. The coast to the E of the hr is low as far as Seaford Hd but the Downs can be seen in the background.
Entrance Within the breakwater keep the W side of hr open in order to clear the mud of the W side between the breakwater and the W pier. The wind baffles under Burrow Hd with heavy squalls in a W blow. Traffic signals from W pier must be obeyed by all vessels. A R ball or G over R lt permits entry or exit by small vessels (under 15m) with care; triangle over ball or G lt, entry no exit; ball over triangle or R lt, exit no entry; ball-triangle-ball or RGR lts, port temporarily closed to all traffic. (Signals other than the first are for commercial shipping.)

Extract 12

RYE HARBOUR 50°56'N 0°47'E. Tel: Camber 0225.

Pilotage: T.H. Compulsory with certain exceptions. E.T.A. 12 h in advance via GPO Coast Station and 2 h in advance via Rye Harbour Radio.

P/Station: Lookout in shore.

Radio—Port: VHF Chan. 16, 14. Hours 1 h before to 1 h after H.W. Listening watch on Chan. 16 for expected vessels.

Entry Signals: Tide Signals from (Day) Flagstaff (Night) Lt. near Rear Ldg. Lts.

By day: only when there is notified movement of cargo vessel either in or out of Harbour. Tidal Balls indicate depth over Bar when vessel due to cross. Lowered when vessel has cleared Harbour or berthed. 2.4m on Bar, 1 ball; 2.7m on Bar, 1 ball on each yard: 3m on Bar, 1 ball at mast-head; 3.3m on Bar, 1 ball at mast-head and 1 on yard; 3.6m on Bar, 1 ball at mast-head and 1 on each yard. By night: less than 2.1m on Bar, no Lt., 2.1-3m on Bar, fixed G. Ltd., over 3m. on Bar, Fixed R. Lt.

RYE FAIRWAY. Lt.By. L.Fl. 10 sec. Spherical R.W. vert. stripes.〰 . Appr. and Anche. By.

Extract 13

RYE   Chart BA 1991

HW +0.00. DS: E −4.00; W +3.00.

(Approach) MHWS 7.7; MHWN 6.0. (Harbour) MHWS 5.3; MHWN 3.6.

The hr should be entered 1–1½ h before HW. The whole hr dries well before LW even at neaps.

Bar The sands off the hr are flat and shoal gradually to the entrance, drying ½M or more offshore at MLWS. The bar shifts in position and height but is of little consequence around HW.

Extract 14

FOLKESTONE   Chart BA 1991

HW −0.15. DS: E −2.00; W +5.00.

MHWS 7.1; MLWS 0.7; MTL 3.9; MHWN 5.7; MLWN 2.0.

The hr entrance dries at LW. Yachts can lie in shelter of the outer pier at any time of tide.

Approach To clear dangers offshore, keep S Foreland open of Dover cliffs by day and the S Foreland Lt open by night. Approach the pier between the bearings of W by N and ENE. The stream sweeps past the outer pier and leaves an eddy. Do not enter if B flag is flying at pier-head as this indicates a ferry is about to enter or leave.

Berthing The harbour is untenable in very strong winds from the E. For refuge in such conditions anch in shingle just W of pier near Trinity Ho Lookout Stn. In normal conditions if not wishing to take the ground anch just outside entrance to fishing hr (to E of pier), sand. Good holding ground and comfortable in winds of up to F6 from S or W. If prepared to take the ground enter the fishing hr from 2h before to 3h after HW and lie alongside quay in thick mud (beware of muddy gulley a few feet out from quayside). Alternatively lie to own ground tackle (fore and aft) among sailing yachts in the fishing hr.

Extract 15

DOVER 51°07′N 1°20′E) Tel: Dover 206560 (P.O.I.S.: 206585)

Tidal Harbour and Inner Docks
   Berthing for yachts in Wellington Dock. Gates open 1 h before HW to HW. Get permission to leave from Dock Master Office open 2½ h before HW.

Extract 16

FRENCH PORTS

Traffic Signals	Full Code	
3 Balls (vert.).	3 R. Lts. (vert.)	Entrance prohibited (emergency).
Ball/Cone (point up)/Ball (vert.)	R.W.R. Lts. (vert.)	Entrance prohibited (normal).
2 Cones point together over Ball (vert.)	G.W.R. Lts. (vert.)	Entrance and departure prohibited.
2 Cones points together over Cone point down (vert.)	G.W.G. Lts. (vert.)	Departure prohibited.
Traffic Signals	**Simplified Code**	
Red flag	Red Lt.	Entrance prohibited.
Green flag	Green Lt.	Departure prohibited.
Red over Green flag	Red over Green Lt.	Entrance and departure prohibited.

Extract 17

BOULOGNE 50°43′N 1°34′E. Tel: 30 10 00 (H.M.) 31 52 43 (Control Tr.)
Pilotage: Compulsory for vessels over 150 tons.
P/Station: Outer Lt. By. or in bad weather. Rade Carnot.
Radio—Port: VHF Chan. 16, 12—continuous.
Radio—Pilots: VHF Chan. 12, 16—continuous when on station.
Entry Signals: Full code shown from W entrance to Darse Sarreg-Bournet for outer Hr. and from SW Jetty, Quai Gambetta, Bassin Loubet, Gare Maritime, Jean Sanson Lock for inner Hr. Special signals alongside full code signals.

Extract 18

CAP GRIS-NEZ 50°52′N, 1°35′E
Lt.Ho. Fl. 5 sec. 27M. white Tr. 72m. Obscured at cliffs of Cap Blanc-Nez and Cap d'Alprech 232°-005°. R.C. Siren 60 sec.

Extract 19
PILOTAGE: CALAIS TO FÉCAMP

Charts Nos. 1892, 2451, 2612, 2613

Both in reaching Calais or Boulogne from England, and in the northern part of the coast covered by this book, the main problem is a man-made one: the traffic separation scheme in the Dover Straits. Crossing the lanes must be done as nearly as possible at right angles, and when proceeding down the coast great care must be taken not to stand so far out to sea as to wander into the NE-going lane, which is only 2.6 miles offshore at its nearest. The lanes are clearly shown on Admiralty Chart No. 1892.

Shoals extend nearly two miles offshore between Caps Blanc-Nez and Gris-Nez, so it is advisable to pass north of the IALA starboard buoys CA3 (Fl. G. 4 s) and CA1: the latter is unlit, so at night substitute the W Cardinal buoy a mile to the W, lit VQ (9) 10 s. From here or CA1, a course can safely be laid to pass a mile W of Cap Gris-Nez, after which keep at least that offing until Boulogne pierhead towers or lights are seen, when they can be steered for.

South of Boulogne the separation zone is soon far offshore, and the only dangers more than a mile offshore are the sandbanks which bulge seawards from the estuaries of the Canche, Authie and Somme. In poor visibility these areas can be tricky, and the careful navigator should keep a good offing while passing each bay, and only close with the coast again when he is sure that he has passed them.

SW of Le Tréport the few hazards are never more than ½ mile offshore, apart from the 1 m wreck NNW of Pte d'Ailly, marked by a N cardinal buoy, VQ. Either pass north of this, or *at least* ½ mile to the south.

It is worth making the point that along the part of the coast that consists of cliffs (i.e. from Le Tréport to Fécamp) the harbours and coastal towns all lie in natural breaks in the cliff line. This has two effects: one is that they all look very similar and are difficult to tell apart, the other that they are often invisible from a boat sailing fairly close in along the coast until they are nearly abeam. It can be quite disconcerting to see a line of apparently unbroken cliffs stretching ahead for miles when your destination is a harbour that the chart says is only a couple of miles ahead, but if your navigation is right it will turn up on cue, appearing like magic out of an apparently smooth cliff.

CALAIS

Charts Nos. 1892, 1352

Calais has been associated with maritime England from time immemorial and indeed it was the last English possession on the French mainland to be lost. There is little doubt now that foreign yachtsmen can consider Calais as an excellent port to visit: good shopping facilities and absolute safety in the yacht harbour, both from the elements and from petty thieving, are amongst its attractions. Furthermore, access and approaches are relatively easy.

Approaches
The harbour entrance to Calais faces roughly NW and there is no difficulty when within half-a-mile. But about a mile N of the entrance Ridens de la

Calais

Charts Nos. 1892, 1352

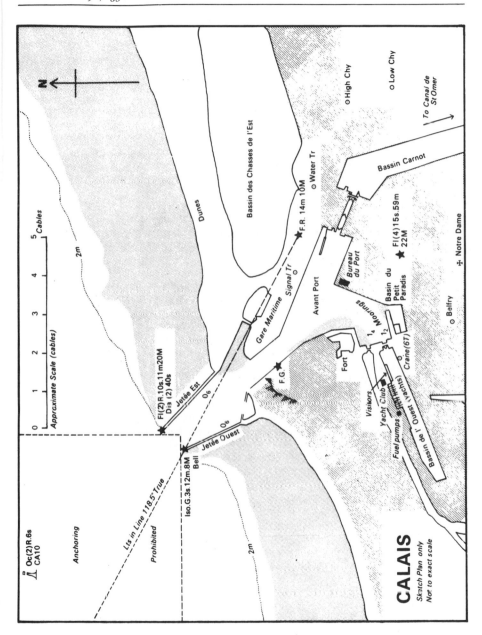

Rade, a sandbank which is long and narrow and which shoals at one place to 0.3 metres is a hazard – in bad weather seas break heavily upon it.

Approach from the E

Although a deep water passage between the sandbank and the shore does exist, it is narrow and unmarked. It is therefore necessary to pass outside the Ridens until the shallowest part has been cleared. The shore should be given a berth of at least two miles until the pierheads bear south, when it is safe to alter course for them, but checking constantly to make sure that any eastgoing tide is counteracted. In calm weather and at half-tide or above, these precautions are of course unnecessary. Conversely, in strong onshore winds, it is wise to hold a course parallel to the shore until buoy CA6 (Oc R 4 s) can be identified and left to port, as the sea can break heavily on the deeper western part of the shoal, even though there is a depth of 3½ metres at L.A.T. At night, remember that the main Calais light (Fl. (4) 15 s), lies nearly a mile to the east of the pierheads, so from the east, course should not be altered until the bearing has decreased to 160° (mag.).

Approach from the W

The red buoys CA8 and CA10 should be left to port and a new course for the harbour entrance should not be made until CA10 is nearly abeam. A glance at Chart No. 1892 will show the configurations quite clearly.

In addition to the foregoing instructions, it is important to allow for the strong tidal stream which sets across the entrance. If arrival has been timed for high water, to allow speedy entry to the inner harbour, the rate can be as much as 3 knots ENEly at springs. As well as the cross-set, strong swirls may be encountered, so the piers should be given a wide berth.

In daytime a simple rule is to keep the lighthouse on the end of the western breakwater in line with the tall chimney (easily visible) when the tide is running to the east. When the tide is running to the west keep the lighthouse on the end of the eastern breakwater in line with the main Calais lighthouse. Traffic signals (shown from the Gare Maritime) are not mandatory on yachts, but ferry traffic is exceptionally heavy and boats equipped with VHF are advised to clear entry and departure with Port Control on Channel 12. In any event a good lookout must be kept for vessels entering or leaving, and yachts must keep well over to their starboard side.

After entering the harbour between the two breakwaters proceed straight ahead to the Avant Port which then divides: these two limbs are at right-angles to each other, the Avant Port de l'Est and the Avant Port de l'Ouest. The eastern limb takes the cross-channel ferries on the northern side and fishing vessels on the southern side. Yachts should not try to moor anywhere in the Avant Port de l'Est but should turn to starboard into the Avant Port de l'Ouest. (It may be noted, however, that the continuation of the Avant Port de l'Est communicates by means of two parallel locks with a basin for small craft, forming the termination of the Canal de Calais, a route to Terneuzen and intermediate ports.) The east wall of the Avant Port de l'Ouest near the harbour office is quite suitable to lie against as a temporary measure, but only for two-thirds of its length. The southernmost third dries, as does the Bassin du Petit Paradis which opens immediately to the south-west. Owners of bilge keelers should resist any temptation to moor alongside the southern part and dry out, as the bottom is dangerous (1982) with jagged rocks and other obstructions. There is a grid in the Bassin du Petit Paradis which may be available for use: inspect it at low water before using. The harbourmaster will advise and make the necessary arrangements.

Most yachts can normally take advantage of the mooring buoys laid outside the lock gates, but it must be stressed that in the whole of the Avant Port, even in calm weather, there is a nasty scend, very often enough to make those with queasy stomachs feel seasick (just like Dover). There are 28 of these mooring buoys, which are intended for yachtsmen waiting for the dock gates to open, or for vessels that have left the Port de Plaisance at high tide, but wish to wait before leaving the port. There is a rule that no buoy may be occupied for more than one tide, but even so, on busy days and particularly at weekends it is often impossible to find a free buoy during the last hour or two before the gates and bridge open, and it is then necessary to moor against the quay wall. The dock gates open 2½ hours before HW, and close 1 hour after or later; approximate bridge opening times are HW −2½, −1 and +1 on weekdays, −2½, −½ and +1½ at weekends. Exact times are posted at the control building at the S end of the bridge. Light signals at the bridge: Amber – bridge will open in ten minutes; Green – pass through; Red – do not pass through. Once in the basin, visitors berth alongside the long pontoon on the starboard side, except boats over 12 metres LOA, who lie alongside the south wall of the basin.

The yacht club (YC du Nord de la France) is large and friendly, with excellent shower and toilet facilities as well as a pleasant bar overlooking the basin. Its telephone number is (21) 34-55-23. Harbour dues, which are moderate for the facilities, are collected by M. Henri Bessodes, harbourmaster of the yacht harbour. His office is on the ground floor of the yacht club, but he is usually on the pontoon for daytime bridge openings to advise on berthing.

Duty-free stores are readily available through the club, on any day except Sunday. Customs office in the Rue Lamy. Tel: (21) 34-75-40. Strangely enough for such a popular harbour there is no chandler, nor repair facilities. Fuel in yacht basin (see Plan 1). Excellent shopping near the yacht basin, and many good restaurants at all price levels.

Tidal information
Height above datum of soundings in metres

High water		Low water	
Mean springs	Mean neaps	Mean springs	Mean neaps
6.9	5.6	0.7	1.8

The ENE stream starts at HW Dover −1¾, and reaches a peak of 2.9 knots (mean springs) at HW Dover. The WSW stream starts at HW Dover +4½, and reaches a peak of 2.5 knots (mean springs) at −5.

The Simplified Method of Tidal Height Prediction Using Tidal Curves from Admiralty Tide Tables

STANDARD PORTS

Example 1:   Finding a Time for a Given Height
At what time (BST) will the tide fall to a height of 3.3m at Dover during the afternoon of 22nd September?

Extract the relevant times and heights of high water and low water from the tide tables (correcting to local time as necessary) and work out the range:

	HW Time	Height	LW Time	Height	Range
Dover	0957 BST	6.3m	1737 BST	1.1m	5.2m (0.3 from springs)

Refer to Extract 1

(1) Fill in the time scale at the bottom of the tidal curve with HW and hours before and after HW as required.
(2) Mark the HW height along the top height scale (H.W.Hts.m) and the LW height along the bottom height scale (L.W.Hts.m).
(3) Draw a range line between the LW mark and the HW mark.
(4) Mark the height of tide required along the top height scale. Draw a vertical line downwards from this point to intersect the range line.
(5) From the point of intersection on the range line draw a horizontal line to the right to cut the rising or falling tidal curve as appropriate. Interpolation between the curves for spring and neap ranges is done visually (where appropriate) after comparing the predicted tidal range with the mean ranges for springs and neaps. (Do not extrapolate outside the spring and neap curves.)
(6) From the point of intersection on the tidal curve, draw a vertical line downwards to intersect the time scale and read off the time, (1353).

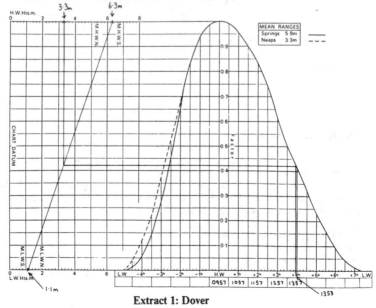

Extract 1: Dover